TIPPING POINT FOR A HOLLOW EARTH:

HOW ASIMOV AND SAGAN BACKED MY PROPHECY OF A POLAR HOLE SHIFT TO EUROPE

BY RUTH LEEDY CARR

DORRANCE
PUBLISHING CO
EST. 1920
PITTSBURGH, PENNSYLVANIA 15238

Dorrance Publishing Co
585 Alpha Drive
Suite 103
Pittsburgh, PA 15238
Visit our website at *www.dorrancebookstore.com*

ISBN: 979-8-89341-289-5
eISBN: 979-8-89341-787-6

TIPPING POINT FOR A HOLLOW EARTH:

HOW ASIMOV AND SAGAN BACKED MY PROPHECY OF A POLAR HOLE SHIFT TO EUROPE

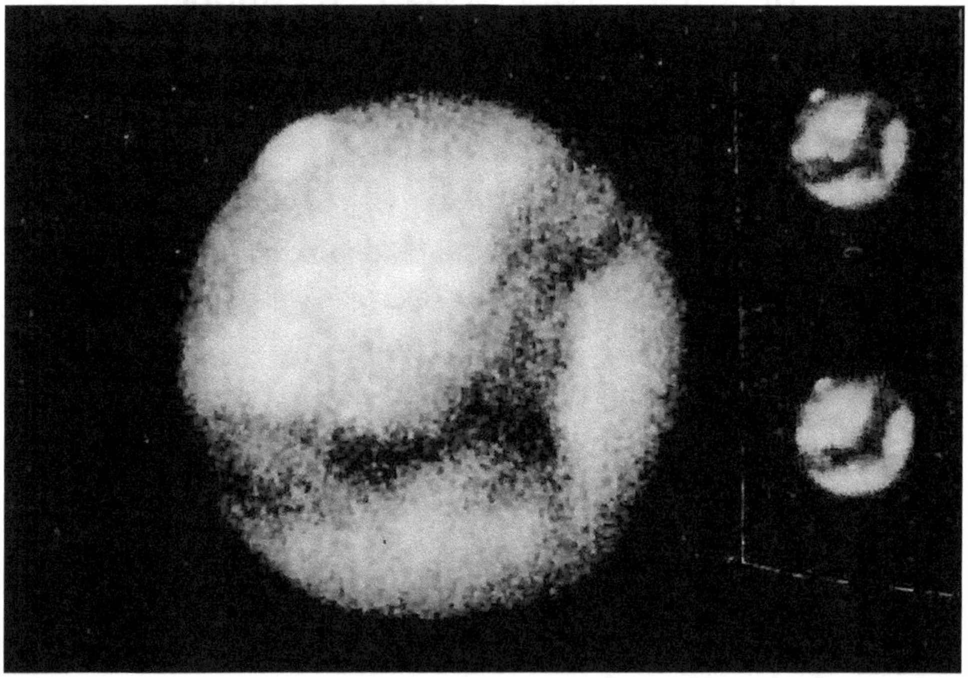

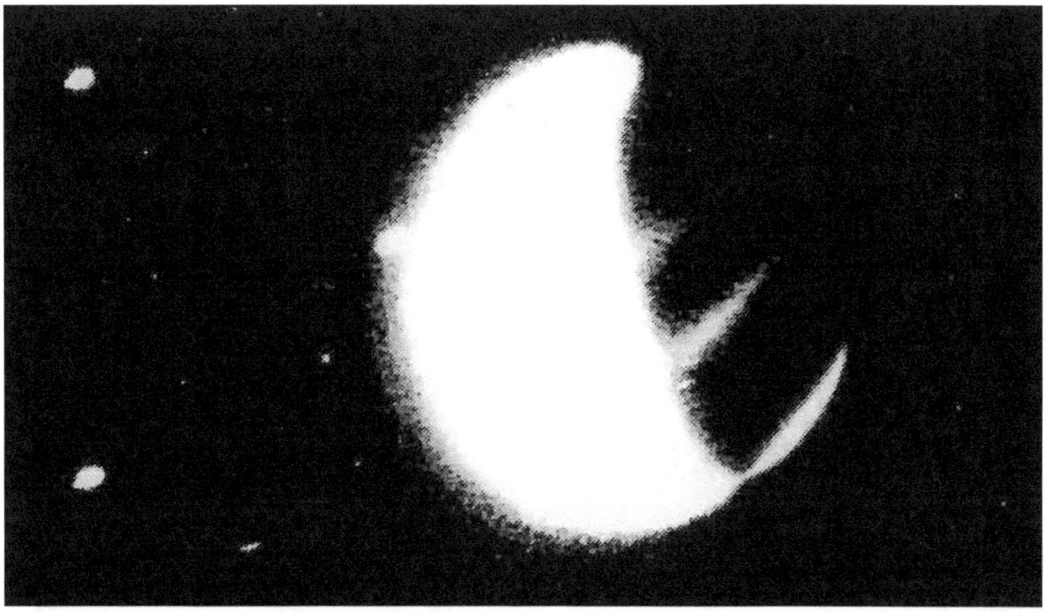

CONTENTS

INTRODUCTION TO HOLLOW PLANET HINTS, RIDDLES, PHOTOS, PLUS CODED WARNINGS OF AN UNSTABLE HOLLOW MARS

Why is a lighted polar "cap" on the night side of Mars a threat to Earth? What do polar hole photographs tell us about Earth's structure and the consequences of a coming axis shift? How did Isaac Asimov and Carl Sagan conspire to blow the lid off the cataclysmic secrecy under which the scientific community has labored? The answers will be found here.

The evidence is clear that an axis shift is on its way, and yet most people have no idea that a geomagnetic field reversal will precipitate an axis shift. They aren't even aware for sure that Earth's axis is able to shift. How can this level of ignorance be prevailing in our time? You will see here how the truth has been systematically covered up, as reported by Asimov and Sagan. They were smart enough to give us covert messages that are clear and unmistakable.

The solar wind's force, which drives Earth's rotation and keeps its outer shell in orbit around the inner sun, no longer drives the rotation of Mars, as will be clearly seen. The axis tilt and loose connection between the axis and magnetic poles will be explained.

If there is <u>no secrecy</u> about the Earth...

1. Why did Isaac Asimov put SOS signs in his "Hollow Earth" essay?
2. Why did Carl Sagan put "field reversal" and "Earth axis shift" into the name of a character warning of doomsday?
3. Why did Asimov put the letters for "Planet Earth be hollow" in only two character names—a mated pair?
4. Why was Edgar Cayce's prophecy for an Earth axis shift dated the same as his prophecy for a special light in the clouds?
5. Why did a geologist request anonymity in saying Cayce's Earth axis shift prophecy <u>could be fulfilled</u> by an approaching geomagnetic field reversal? Why did he say nothing about the prophecy of a special light in the clouds in relation to this?

6. Why did the person speaking through Cayce during trance readings choose to remain anonymous?
7. Why did NASA refuse a congressman's request to comment on a diagram showing polar light to be on the night side of Mars, while freely listing non-photographic evidence for a solid planet Mars with no internal sun visible through polar holes?
8. Why did Asimov secretly use three names from his novel about a secret star threatening Earth as a coded message? We find the names Merry, Tessa and Ranay in the letters next to commas and apostrophes of another story. Why the secret message?
9. Why did a *Scientific American* article giving details of an approaching reversal of Earth's magnetic field fail to mention that this is considered to be a cataclysm, associated with extinction events?
10. Why did Carl Sagan put a "dots and dashes" reference inside an SOS sign consisting of real dots and dashes?
11. Why did Isaac Asimov create three SOS signs using actual dots in a story where SOS signs are actually sent and Earth is actually threatened?
12. Why would Asimov claim to be a debunker of wild theories and yet encourage speculation by writing that a thunderstorm's noise "rattled across a hollow world"?
13. Why did Asimov use hollow planet imagery three times in the 1980 interview where he predicted that civilization would not survive until 2010?

Most of the riddles you will find here are ambiguous, because sending a clear, definite message is much more dangerous when a cover-up is powerful. But some of the riddles are <u>not</u> ambiguous. When Isaac Asimov's name can be found in a short riddle, that is not an accident. He put my name in his riddles as well, since I was writing about him. He also gathered information about me, to be put in riddles so I could use it later to show that he endorsed my work.

"I am both looking and listening." Asimov used the character Sam Chase to help convey that he had witnessed my landlord Sam Wilson chasing me on two occasions. When my landlord C.T. Tincher proclaimed "I am a sinner," and told me he was teaching me to be a better lover, Asimov was paying attention. My piano playing, my father's poetry, and my brother's dramatic name change were also among the details used by Asimov and Sagan to define me as someone whose work they endorsed. Asimov even arranged twice for agents to contact me and talk about him.

"I've been listening to U, R.A. Leedi. I believe U are going to be our savior." This is from "Stor Gendibal, Sura Novi." In Gendibal's name alone there is this:

"I be listening to R.A. Leedi." Gendibal's colleague "Lestim Gianni" is a perfect anagram of:

"I am listening." The month of "Theptar," when civilization is predicted to fall in *Nightfall* carries Asimov's message for us about civilization's end:

"Prepare, ere the Earth teareth apart."

1.
SIGNS OF THE COMING FATE OF NORTHERN EUROPE, JAPAN, HAWAII, CALIFORNIA, AND THE PLANET MARS

"There will be trouble" over Edgar Cayce's cataclysmic revelations. This prophecy, which was included in a trance reading by the early twentieth century medium Edgar Cayce, tells us that his prophecies relate to secrecy about coming cataclysmic events, not just the events themselves. And so when the prophecies are veiled, especially regarding "His Light" in the clouds, we can assume these are the most sensitive, secret matters that are being alluded to.

"There will be trouble" when the truth comes out (149). It will be a scandal. The complex method by which these prophecies were delivered seems to have been designed to prevent them from being suppressed, as we shall see. (Afterword gives more detail.)

Why should we believe Cayce, or listen to him at all? There are several reasons.

1. A geologist confirmed that the axis shift prophecy is believable because an approaching reversal in Earth's magnetic field is likely to result in an Earth axis shift (149). (See Bloxam and Gubbins for more on geomagnetic field changes.)
2. Carl Sagan included "Earth axis shift" as well as "field reversal" in the name "Stefan Alexeivich Baruda" in his novel *Contact*. Baruda issues a fiery doomsday warning at a time when alien contact has already roused fears in the public mind that doomsday or even an Earth axis shift may be imminent. Baruda's warning tells us that "Earth axis shift" and "field reversal" did not get into his name by accident. Carl Sagan put them there as a warning to us. From the same novel, the name of the prisoner "Tyrone Free" can be unscrambled to read:

   "Eye R not free." Sagan's book title *Broca's Brain*, sounding like "broke his brain," suggests a kind of mental bondage preventing Sagan and other experts from spilling cataclysmic secrets.
3. Isaac Asimov included many references to Earth axis shifting and axis migration in his riddles.

His riddle messages relating to northern Europe, Japan and California conform to the prophecies of Edgar Cayce. I believe he even referred to Cayce in a riddle. Asimov also placed unmistakable secret messages in the letters adjacent to punctuation in his final works of fiction. His covert communication also included SOS signs in his essay "The Hollow Earth" (64) and elsewhere, and various hollow planet riddles.

4. Cayce is worth our attention because his prophecies paint a consistent picture of changes caused by an axis shift, including a new polar hole in northern Europe and a new equatorial bulge near Japan and the western United States, as well as the appearance of land off the U.S. east coast and the advent of warmer climate in Alaska.
5. There are accounts from explorers that support Cayce's implicit message as to the existence of polar holes.

Someone else's voice spoke through Cayce during his trance readings. At one time this would have sounded like a mysterious and magical phenomenon. Now, however, we are learning enough about brain science and technology to understand that there was a mundane explanation for all of this. Some sort of chip was evidently placed in his brain enabling an outside actor to beam signals activating his brain and causing him to speak. The outside actor possessed this technology by virtue of belonging to an advanced society based (most likely) in Earth's interior.

"PLANET EARTH BE HOLLOW" – ASIMOV'S DOUBLE MESSAGE

Asimov's hollow planet riddles included easy messages like "Mars be hollow" from "Hober Mallows" (34) and "We have a hollow world" from "Hor Devallow" (65), as well as longer ones like this from "Henry J. Lafkowitz" (91): "Laff if yew want to, yet I know the hollow Earth theory iz no joke."

In spite of many examples like this in Asimov's fiction, I have found the words "Planet Earth be hollow" in only two of his character names, and these were a mated pair in the same story. The message "Planet Earth is hollow" can also be found in one of these names.

"Hortense Hepzibah Lowot" from "Perfectly Formal" (45) is the name containing both "Planet Earth be hollow" and "Planet Earth is hollow." The letters for "hollow" are grouped together near the end of her name. She is called "frontally concave," reminding us of the concave inner surface of a hollow planetary shell orbiting its central star, a structure also found in the spheroidal planetary nebula (132).

"Winthrop Carver Cabwell" is engaged to Hortense in the story. Both these names contain longer messages than "Planet Earth be hollow," and my attempts to pin these down will be seen later. The double statement "Planet Earth be hollow" in the names of Winthrop and his fiance is a more definite and significant statement than any longer message we can find in these names.

Both an Earth axis shift and the collapse of a hollow Mars will be seen here as coming events which were of concern to Isaac Asimov and Carl Sagan. The way that the solar wind defines and redefines our planet's axis in addition to powering rotation will be examined here in detail. Since Mars no longer has north and south magnetic poles to serve as target points for this powerful solar force, the rotation of Mars cannot be receiving any solar wind support which in the past kept its hollow shell spinning fast enough to remain in orbit around the central star. The resulting loss of rotation speed leading to the collapse of Mars is a major worry for me, as it evidently was for Isaac Asimov and Carl Sagan. (See Gribbin and Plagemann regarding the solar wind's influence on Earth's speed of rotation.)

A STAR FROM MARS HEADING FOR EARTH?

Earth is threatened by a secret star in Asimov's 1989 novel *Nemesis*. Space people who know about this star have kept its existence secret from the people of Earth. It sounds like nothing more than pure science fiction. And yet Asimov lets us know in dramatic fashion that there is a secret star destined to head for Earth, and he shows us where that star is.

Asimov used a special code that is 100 percent reliable (unlike puzzles) to convey this warning to us. I went looking for a reliable code in Asimov's work after many years of seeing that riddle messages from Asimov and Sagan which I reported in my newsletters were being ignored. People evidently thought I was imagining the existence of riddles in the work of these experts. So I decided to look for a definite type of message in the letters adjacent to punctuation in Asimov's many stories. Writing down the letters before and after every period, or every comma and apostrophe, in a section of text, was very tedious.

I did hit pay dirt, but didn't have the patience to do a very wide-ranging search. I may have missed many fabulous messages, but here is the best one I found:

The names "Merry," "Tessa," and "Ranay" jumped out at me from an otherwise random list of letters adjacent to commas and apostrophes in Asimov's story "Frustration" (40). These were three female character names from Asimov's novel *Nemesis*. It is quite rare to find any five-letter word in a list of letters adjacent to punctuation. Therefore, finding three names of that length, all from the same novel, had to be the result of a deliberate effort by Asimov to place these names before the public. What was he trying to tell us? At first I was disappointed, thinking it was just another riddle. But no, he was using the three names to direct our attention to the novel *Nemesis*.

"Read my book about the secret star." This was the message conveyed by the three names. As to how Asimov indicated that Mars had to be the source of the secret star, the name of the book's main character or "star" —Marlene—is an important clue. Asimov's message here was obviously cryptic. Clearly he didn't want it to be extremely obvious that he was sending a coded message, nor did he want the message to be extremely obvious. This was evidently for his own safety and the safety of his editors and publishers. This was an extremely explosive message, a real hot potato, and it had to be handled gingerly.

"Lest we al die, die, die," which was another possible message of this type, turned up adjacent to periods in the section of Asimov's final novel (30) where the character "Tejan Popjens Lih" is introduced. In that name, we find obvious anagrams of "Janet Jeppson," the name of Asimov's wife. The full message in that name appears to be: "I hope Janet Jeppson sees people, is not all alone." Asimov died of cancer in 1992 at age seventy-two. Various other riddles about Janet Jeppson and Asimov himself will be seen here as we go along. We will see Asimov declaring his ambition to become a major science writer. We will see the astronomer Carl Sagan declaring "I peer at every planet" —an anagram message taken, naturally, from the name of an astronomer.

LIGHT IN CLOUDS DURING EARTH AXIS SHIFT?

The most cryptic and explosive prophecy from Edgar Cayce is that the forty-year period ending in 1998 "will be hailed as the time when His Light will be seen again in the clouds." This sounds mysterious, and since the year 1998 has long ago passed us by, we may be tempted to dismiss the prophecy as some kind of a mistake. But this prophecy can be understood, and can be interpreted not to be a mistake at all.

The shifting of Earth's axis was predicted by Cayce to occur in the year 2000, and was predicted to cause dramatic changes in climate for various parts of the world. This date corresponds closely to the 1998 date predicted for the appearance of "His Light" in the clouds. The question then becomes, why should an Earth axis shift result in the appearance of light in the clouds? This question is answered by Marshall B. Gardner's 1920 book as he explains that Earth's inner sun shines through the north polar opening and is reflected in the "water sky" that is typical of the Arctic region: On reading this I had a flash of insight into Second Coming prophecy. This was the key. A light shining up from ground level would have to be reflected back down in order to be seen clearly. That is why clouds had to be mentioned.

For this to be a Second Coming prophecy we would have to conclude that Jesus considered the Earth to be hollow and portrayed himself as a symbolic representative of the inner planetary sun shining through polar openings newly recarved by centrifugal force in an axis shift. (Orbiting matter must remain a certain distance from the axis, as Gardner points out.) The case can be made that Christ did intend this meaning, as we shall see.

CAYCE PROPHECIES INCORRECTLY DATED ON PURPOSE?

Why did Cayce give us an incorrect date for the cataclysm? In my own research I was motivated by the deadline of the year 2000. If the deadline had been set for the year 2050, for example, I might have thought there was plenty of time for others to figure out what the prophecies meant and share that with the public. The speaker behind Cayce's prophecies may have deliberately set the deadline too early, realizing that the cataclysmic project would take many long years to achieve success, given the strength of the cover-up.

The light-in-clouds prophecy had to be cryptic and mysterious to ensure the safety of the church officials and scholars who had to convey this gift to future generations.

LIGHTED POLAR HOLE MISIDENTIFIED AS ICE IN MARS PHOTOS

There are various Mariner 7 photos of Mars taken on August 4, 1969 (145, 143, 132, 137) that show a lighted polar "cap" on the night side, which in reality must be a lighted polar hole. Several of these photos show clear evidence of retouching to conceal the day-night border at the right edge of the polar "cap." Other Mars photos show protruding blobs of polar light which are sometimes described as ice caps, though later satellite data reveals no ice cap at all— only wet rocks. These pictures can be seen in Chapter 2. The key point about the lighting of Mars is that it must follow the same rules that apply to the Moon, if we assume only one light source. The lighting of Mars, if done solely by the sun, must show the planet going through phases like the phases of the Moon, with only 180 degrees of the perimeter visible at one time.

In the Mars photos of August 4, 1969 the lighted polar "cap," which is very close to the perimeter, clearly violates this 180-degree rule, and so this light must be shining from within through a polar hole. In the Gamow and Moffat-Schneour photos there are startling blobs of polar light extending well beyond the surface. Dismissing these as polar ice caps was absurd, since a reflection from ice at a low sun angle would be unlikely to produce such a dramatic light, and the authors had no evidence for the existence of any ice on Mars.

"Mars shall end and its inner star shall threaten the Earth." This message can be found in "the Marshal Nedelin" from Carl Sagan's novel (148). Isaac Asimov provided numerous similar messages.

"Ruth A. Leedy, you say the Earth is hollo. You are right." This message can be found in "Agatha Dorothy Lissauer" from Asimov's story "The Critic on the Hearth." (Lissauer, like me, was a writer who had a boyfriend but no definite plans for marriage.) Since "Earth" is shown in this title, we can surmise a message may be there, as follows: "I cannot teach the correct theori on the Earth. Rooth Anne can."

Many messages along this line can be found. The most convincing ones are the ones where Asimov's own name is given in the riddle, sometimes in a very obvious way. The letters for "Azimov," the original spelling of Asimov's name, can be found together in the name of a wise planetary leader in Asimov's 1982 novel (34). In "Endomandiovizamarondeyaso" we can find this message: "I, Asimov, say Mars is doomed and may end in seven-zero or nine-zero years." I didn't use the "z" to spell "Azimov" because that spelling is no longer correct. Asimov did give to *Mother Earth News* a dated prediction for the end of civilization (144), and his 1980 remarks included what I took to be three hollow Earth hints. The magazine cover asked: "Thirty Years to Armageddon?"

"Eye worry over events that threaten to shake or even destroy the Earth seventy years down the road." This message can be found in "seventy dworshaks" (51), where Asimov has clearly indicated the words "seventy" and "shake."

"Upper Europe shall end soon, as you and Ed Cayce prophesy, R.A. Leedy." This message can be found in "Euphrosyne Durando, Cal" (38). Cal is a writer using the pseudonym Euphrosyne Durando. The phrase "under a hundred years" could be part of this message.

Edgar Cayce's prophecy regarding Europe is that "the upper portion of Europe will be changed as in the twinkling of an eye."

This suggests a new axis in upper Europe, which is consistent with other prophecies.

Cayce also predicted that "the greater portion of Japan must go into the sea." With a new pole in what is now northern Europe, the equatorial bulge would fall near Japan, Hawaii and the United States west coast, resulting in higher water levels in all these areas.

"I want to warn Japan, Hawaii. That won't happn." This message can be found in "Jonathan Winthrop" (29). In "Francis Rodano," his partner's name, we see: "Franc and San Francisco no ar saf." In "Finangelos" (30) there is this: "Los Angeles no is safe." In "Jander Panel!" (114) we find: "All Japan land endanjered."

The name Euphrosyne suggests "Europe" to us, and "north" can be found in "J.F. Northrop," who has created a character named "Euphrosyne Durando" in Asimov's story "Cal" (38). The two names together make for a long riddle, but after "northern Europe" is removed, the riddle is not so long. This message can be found in the two names: "A dysaster's to destroy Japan and just the northern part of Europe." This respelling of "disaster" is justified by a word like "dystopia." In "Euphrosyne Stump Mellon, Alexius" (50) we find: "Ruth Anne L., you say millions o' people must exit northern Europe prior to an Earth axis shipht. This is true." In "Oola, the All-Purpose Pet" (119) we can find: "The hollo Earth's upper polar hole shall tear apart upper Europe." The word "shatter" could also be used.

TWO DATED AXIS SHIFT PREDICTIONS

Carl Sagan may have included a dated prediction for an Earth axis shift in the name of "Stefan Alexeivich Baruda" (148), whose doomsday warning was mentioned earlier. The full message in this name may be: "A field reversal shall cause the Earth's axis tu be shifted bi the sun in five tu seven decades." The establishment of new magnetic polar sites in opposite hemispheres from where they are now would redirect or refunnel the solar wind, causing it to define an axis at those new sites.

"The Earth's axis shall be shifted by the sun in fifty tu a hundred years." This message is from "Synnax, the Blue Drift" in Isaac Asimov's novel *Foundation*, as he describes the location of the home world of the character Gaal Dornick. Confirmation of this message is provided by repeated use of the word "axis," sometimes buried in the word "taxis," in nearby pages.

"Axis migration is near." This message can be found in "Santirix Gremionis" (114). Asimov confirms this message by having Gremionis ride on a scooter, each of whose wheels turns on an axis that migrates as the scooter moves along.

"Sometime soon the axis shall shift and millions shall die." This message can be found in "flaxated his modinem" (60). Asimov does not portray an axis shift in this story, but there are hollow planet hints. Earth is called a "hellhole" and a "pure sun of sobriety" is mentioned—a sun that is separate and distinct from the sun we usually see in our sky.

"You'll soon see nu poles in nu positions." This axis shift prediction can be found in "Louis Peyton" (105) in one of the Dr. <u>Urth</u> mysteries.

Earth's rulers apparently are great apes—the Sasquatch—as will be shown when we take a close look at Asimov's covert statements about Earth's ruler in Chapter 10. As a preview of that, we can see in Asimov's story "The Monkey's Finger" (104), about a super-intelligent monkey, that two character names can each spell a version of the word "monkey." These names are Marmaduke Tallinn and Lemuel Hoskins.

The Sasquatch might have a couple of motives for wanting us largely or completely wiped out. It would be natural enough for our simian cousins to prey on us as a matter of policy, if they could get away with it. That would be no different from our policy toward many living creatures, namely, to prey on them insofar as possible without wiping them out. If those running the planet from within are planning to colonize other worlds, they might want us eliminated as a precaution against being attacked by us once we are released from their close supervision and control.

PUZZLES ABOUT PUZZLES

The word "puzzle" is used in Carl Sagan's novel in commenting on the presence of Prebula of Switzerland at a conference. Prebula's presence may be a puzzle, but we also note that "Prebula, Switzerland" contains all the letters needed to spell the word "puzzle." Thanks to Sagan's broad hint that this is a puzzle about puzzles, we can find this message:

"Brain is disabled bi ruler, sew I write in puzzles." The letters for "brain" occur together at the beginning of Sagan's character name "Ian Broderick," where we can also find "broken." The complete message here may be:

"Brain can be broken and I cood die. A code-breaker be needed." This reminds us of Sagan's book title sounding like "broke his brain." If the name of Broderick's countryman "Annunziata" is added to the puzzle, the solution may be: "I can't be direct becauze brain can be broken. But I can uze a code." Many messages similar to this can be found.

"In a puzzle I can tell all." This is from "zapulniclate" (19), a word used by Asimov's fictional demon Azazel. In "a dear zapulnik" (51) Asimov gives us "zapul" <u>again</u>, impressing on our minds this letter combination which strongly suggests "a puzzl." The message here appears to be: "I kan read nearli ani puzzle."

Another puzzle about puzzles from Asimov comes from "zplatchnik, blue cord" (45). Azazel decorates his tail with a blue cord which he calls a zplatchnik. In these words we find: "Pleaze decipher the puzzles that Isaac A. put in each book he rote." One could add my name here, spelled Ruth Anne Leedi. A similar message appears in "Brinsley Sheridan Cooper" (107), where we find: "Brain is disabled. I need your help. Please decipher and share riddles." Likewise, in "Genovus Pandaral" (79) we find: "U all are endangered. Please solve and share pusles." In "Theophilus" (18) we see: "I hope U'll see these pusles." As of 1976 Asimov's appeals along this line were apparently going unanswered. In "Nellie Griswold" (62) we find: "End o' world riddles need reeding. No one is willing."

This is from a novel where three character names begin with "Mar" and one even begins with "Mars." In "Anthony Marsogliani" we find: "I am not only hinting that Irth is hollo. I am also hinting that Mars is hollo." This message, with improved spelling, can also be found in "Montgomery Harlow

Stein" (83): "I am not only hinting that the Earth is hollow. I am also hinting that Mars is hollow." His fellow character "Neville Preston," who is a judge, may provide this message: "I'll never rest till I've proven it."

ASIMOV USES SUPER-RIDDLE LINKED TO BATMAN

Earlier we saw two of Asimov's character names which both could spell "Planet Earth be hollow." In his final mystery (53) he alludes to Batman's valet "Alfred Pennyworth." This is another super-riddle, and Batman's fellow character the Riddler seems to be a hint that an important riddle is associated with Batman. Whether Asimov composed this riddle or not, the fact that he used it shows that he endorsed its message. In "Alfred Pennyworth" we can find "hollow planet Earth" as well as "world" and "theory." The full message may be: "Eye don't dare openly defend the theory that Earth and another planet or two are hollow, for fear of the wrath of the Lord of the World."

I sent Asimov five of my books between 1981 and 1988. One of them, with his name in the title, inspired him to write an essay entitled "The Hollow Earth" which contained SOS signs, as discussed in Chapter 4. My books may also have inspired him to create this anagram message: "It isn't easy to quiz anyone about you." This is from "Sotayn Quintesetz ABT" (34). We can also find messages indicating that my project on hollow Earth and polar shift was potentially hazardous to my health.

"Why are yew riscing yewr life, R.A. Leedy?" This question can be found in the names of two agents for the government of Earth in Asimov's novel *Nemesis*. They are "Garand Wyler" and "Crile Fisher." Fisher's undercover assignment would result in his execution if he were discovered. In his name alone we can find: "R.L. rises her life." The word "feerlessli" could be added. I knew that there was a risk in opposing a powerful cover-up which others were unable to defeat. But it had to be safer for someone like me with no science degree—I am a journalism graduate of Indiana University, 1968—and not much money to put out homemade books than for someone like Asimov or Sagan to try to expose cataclysmic secrets. I must admit my publisher died after republishing my books of 1981 and 1983. Gray Barker of Saucerian Press had been the original publisher of Raymond Bernard's classic *The Hollow Earth*. Bernard himself disappeared after producing this one book that was republished in hardcover and paperback. He was rumored to have settled in the Earth's interior. But perhaps he was silenced because he had been too successful with his message, and so I may have come close to being silenced myself.

"R.A. Leedy, yoo fearlessly warn world of danger." This message is from "Farley Gordon Wells" (119) in a robot novel written by Asimov and his wife Janet.

The use of anagrams to carry messages is recognized by Asimov in his story "Irrelevance!" (26). Two characters discuss the possibility that "Tar Heel, North Carolina" may be an anagram, and a possible solution is offered. When we notice that the first five letters of this can spell "Earth," we should have no doubt that it <u>is</u> a puzzle. The message may be: "I cannot clearli tell the trooth on the hollo Earth theori."

WERE POLAR EXPLORERS CONSPIRING TO LIE?

Accounts of explorations within the Earth or sightings of the inner sun are found in several sources (126, 130, 131, 134). I like the modest report of Danish explorer Fridtjof Nansen (134). No one could accuse him of grandstanding, since he concluded that the dull-red sun he and his crew had seen during the polar winter night was nothing more than a "mirage of the sun." This was merely written in his diary, not announced to the world in dramatic fashion.

If the explorers have all been lying in their claims of seeing Earth's inner sun or other signs of a polar hole (151), or even visiting the interior, one has to wonder what would motivate them to conspire to lie in this way. Also, how could the conspiracy be coordinated and held together over a span of many decades?

Ordinarily when four to ten people see the same thing, such as a grizzly bear in the woods, we do not assume the reports are false simply because other people have failed to notice that phenomenon. Assuming that witnesses are wrong could have grave consequences, and this is nowhere more true than it is regarding reports that the Earth is hollow. If the inhabitants of the interior had the means to keep this a secret, the chances are very good that they would do so. The motivation to lie is much stronger in the case of falsely claiming Earth is solid than it is for falsely claiming it is hollow.

There is plenty of evidence for a hollow Earth, as we will see. The photographic evidence for a hollow Mars is strong, and this alone goes a long way toward proving that our own planet is hollow.

2.
THE PICTURES THAT SHOUT OUT THAT MARS HAS POLAR HOLES LIGHTED FROM WITHIN, LIKEWISE EARTH AND VENUS

My congressman was impressed enough by my newsletter about polar light on the night side of Mars that he forwarded it to NASA in 2001 for their comment. Representative Wayne T. Gilchrest of Maryland had been a teacher of mathematics prior to running for Congress. My attempt to locate the sun's zenith point on a photograph of Mars must have looked like an intriguing problem in geometry to him.

The zenith point, of course, is the center of the day side, the point above which the sun's light is coming straight down to the surface. The day side should extend for 90 degrees in all directions from the zenith point, plus a few degrees of twilight or albedo, which is a reflection from the atmosphere. When the Moon goes through phases, half of the perimeter, or 180 degrees, is illuminated, unless the Moon is full or completely in shadow. The 180-degree rule also applies when Mars is going through phases. The Mariner 7 probe, on August 4, 1968, was able to get a good view of Mars in various phases as it passed from the day side to the night side at a distance of almost 300,000 miles.

Four views of Mars in various phases are presented here in figures 1 through 4. My newsletter which was sent to Rep. Gilchrest was based on Figure 2, also referred to as Photo B in a four-photo sequence. Photo B was the first of these four photos that I discovered. My first husband Michael Callinan was with me in a library in 1981 as I was searching through science books in preparation for my next book on a possible hollow planet catastrophe. Michael set a book in front of me open to a full-page photo of Mars taken on August 4, 1969.

I was stunned to see a bright south polar "cap," as it was called, surrounded by an immense expanse of darkness. The rest of the southern hemisphere was almost totally dark. The polar "cap" was more solidly bright than any other part of the planet. I had already read Marshall B. Gardner's comments about a strange looking polar light on Mars which he believed was shining from an interior sun through polar openings into the sky. But even without that background, I like to think I would have been shocked by this large Mars photo in Patrick Moore's *Guide to Mars*. (This picture was omitted from a later *Guide to Mars* by the same author.)

I wanted to know where the center of the day side was, so I could find out whether the day-night border would place the polar "cap" on the day side or on the night side. The Mariner 7 camera was located high above the center point of the image, as it would look if fully illuminated. This point had to be well below the equator, based on the position of the landmark Olympus Mons. My diagram showed the Mariner 7 camera hovering below the equator and recording a bright solar reflection which was located slightly above the equator. Tracing that reflection back up to its source, I had to place the sun's zenith point somewhat farther north than the bright spot of reflected light. My calculations were imprecise, but that didn't matter. Even if the sun had been hovering directly above the equator itself, half of the polar illumination would have been unexplained, because the day side could only have extended as far as the axis.

The sun was clearly located above a zenith point that was north of the equator. The picture was oriented in such a way as to portray the entire solar system as tilted. To perceive the picture in its true light, one should rotate it clockwise until the bright spot of reflected light is halfway up. By doing this we portray the sun's light as coming in horizontally, and the solar system is level. When we make this change, the polar "cap" is now facing to the left, away from the sun. It is now clear that the day-night border is located at the top and bottom of the image (when the reflection is halfway up). Therefore, only a small sliver of light on the right edge of the polar "cap" can be receiving the sun's light. The rest of the "cap" must be on the night side, which means the entire "cap" is a polar hole illuminated from within the planet.

If the day-night border had been clearly shown descending to the right edge of the polar "cap," it would have been obvious that that "cap" was on the night side. By artificially darkening the southern hemisphere, the photo retoucher was able to confuse the viewer as to how far the day side naturally could extend.

I do not blame NASA or other astronomers for this deception. They did what had to be done to present a controversial photograph to the public. The picture, even with alteration, is a strong statement to the effect that Mars is indeed hollow.

We may be inclined to forgive Patrick Moore for displaying Mars with its axis upright. Turning to Photo A in Figure 1, it is not so easy to justify the tilting of Mars which we see here. The image is tilted so that the polar "cap" is facing to the right, and sunlight is coming down from even farther up than shown in Photo B. This absurd tilt appears to be designed to fool the viewer into thinking the polar "cap" is on the same side, the day side, as most of the illuminated portions of the northern hemisphere. Here again, the lighting pattern becomes clear when we turn the image clockwise so that sunlight is striking the planet horizontally. It is then clear that the polar "cap" is facing away from the sun and must be on the night side. Once again, there is artificial darkening of nearly the entire southern hemisphere (extremely unrealistic), which leaves us confused as to where the day-night border is in relation to the polar "cap."

For sheer variety of retouching techniques, Photo C exceeds all the other photos in this series. Before we examine the changes made to this image, let us get back to Rep. Gilchrest and see what kind of response he received from NASA.

The NASA response to Rep. Gilchrest was dated August 17, 2001 and bore the code L:NKM:mtg:L/2001-486f. The letter was signed illegibly (the last name may have been Rothman) by someone writing for Jeff M. Bingham, Associate Administrator for Legislative Affairs (Acting). This was someone, in other words, whose job it was to keep members of Congress in line, tethered to reality, as it were. How could NASA squelch the issue I had raised when the evidence was clearly on my side? No problem. Once again, I don't blame NASA for this. They did what they had to do.

NASA had an easy excuse for not commenting on the Mars night-side light issue in any way. All they had to do was characterize my newsletter as propaganda and politely ignore the genuine curiosity of a member of Congress who did not have any ax to grind, none whatsoever.

The NASA letter began: "Thank you for your letter of June 18, 2001, on behalf of Ms. Ruth A. Carr, who has submitted information to support a theory that Mars is hollow, with an internal sun."

Here at the outset was a total putdown. I wasn't a concerned citizen raising an important issue about the lighting of Mars. I was merely a promoter of a peculiar theory. As for Rep. Gilchrest, his legitimate curiosity was totally swept aside as unimportant, in spite of his position.

The letter went on to say that NASA does not comment on "speculative lay concepts" submitted by members of the public, and my work should be submitted for publication in a scientific journal. All this was said with such pomposity that I do not care to repeat it in full.

NASA was wrong to try to shunt all responsibility for the Mars night-side polar light issue to some scientific journal. The existence of polar openings on Mars has immense implications for more than one type of cataclysm. It is up to the government to break this news to the public as gently as possible and assure people that the threats involved will be dealt with.

The letter mentioned various kinds of non-photographic data that NASA has collected on Mars, such as mass and moment of inertia, and it was stated that "all the data" supports the solid planet model with regard to Mars. To use the phrase "all the data" while pointedly excluding photographic data from the list was highly misleading. In astronomy there is no kind of data that is more important than photographic data.

The weight of Mars is so great that it could not possibly be hollow, the letter said. This ignored the concept of a hollow shell being in <u>orbit</u> around the central star and held in place by centrifugal force. This argument about the weight of Mars was irrelevant because my work was not focused on proving Mars to be hollow. It was focused on proving that the lighted polar "cap" was on the night side. Rather than confront the central issue here, NASA was doing an end run around the issue, trying to defeat it by changing the subject.

A HIGHLY RETOUCHED PHOTOGRAPH OF MARS

The small version of Photo C which appeared in the *Encyclopedia Britannica* (1974) looked so unnatural to me when I found it that it seemed to be more like a painting. The image of Mars was ridiculously small, under two inches in diameter. Fortunately, the best enlargement I produced revealed important details about the lighting and the retouching that had been done. The entire image had a washed-out, overexposed quality, which caused practically all surface features to be obscured. The landmark Olympus Mons could barely be made out. One could see that the background sky had been darkened over most of the southern hemisphere, even in areas where night-side darkness should have made it impossible to see the outline of the image. In other words, the bright polar "cap" was artificially made to appear as though it were on the day side of the planet, when this was physically impossible. The sun's bright reflection on the side of the planet can be made out in spite of the general lightness of the print. When we rotate the image clockwise so that sunlight is coming in horizontally, it is obvious that the day side can extend only to the top and bottom of the image, and the bright polar "cap" is tilted away from the sun. The day-night border at the right edge of the polar "cap" has been obscured by gray paint, which extends across most of the southern hemisphere, even far to the left of the polar "cap." This gray paint, which is contrasted with the artificially darkened background sky, causes the perimeter of the image to be visible for well over the 180-degree limit for a planet that is approximately in half phase. No planetary image lit solely by the sun can have a perimeter that is visible for 240 or more degrees around the edge. Anyone who has studied the phases of the Moon is aware of this. In short, the picture is a fraud. Its original appearance prior to retouching may have been similar to the sketch in Figure 5.

This Photo C image (Figure 3) has been retouched in such an obvious manner that it is really a gift to us. The glitch in the perimeter, halfway up on the right side, shows us clearly where the sharp artificial darkening of the background sky was initiated. The overexposure of the photographic print is so extreme as to be obvious to almost anyone. Even the polar "cap" appears to have been artificially smeared, when compared to its appearance in photos A and B. Smeared polar caps can also be found on two Mars photos in *Pictorial Guide to the Planets* (137), where the "ragged melting edge" of the supposed ice cap is described in the caption. No water had been proven to exist on Mars when this book was published. Even today, only wet rocks have been detected near the north pole. And so the supposed ice cap on Mars is a myth, pure and simple.

The experts were probably all aware that statements about Mars polar ice were fiction, but the lie had to be told in deference to a powerful cover-up. The lie is so flagrant that truly it is a gift to us, just as all these pictures are gifts, if we are willing to study them.

In Photo D (Figure 4) the bright polar "cap" has been partially cropped off. This picture is from the title page of *Pictorial Guide to the Planets* (137). This cropping should strike us as odd, since the polar "cap" is far more solidly bright than any other part of the image and since it is difficult to explain how the sun's reflection, at a low angle, could produce such brightness. In other words, the most spectacular and inexplicable element of the photograph has been partially covered up. The background sky has been artificially darkened underneath the polar "cap," evidently to give the impression that that "cap" is on the day side. The upper portion of the image appears to have been painted so as to extend the day side to the left (so the polar "cap" doesn't look so lonely in its brightness). The day-night border at the right edge of the polar "cap" has been obscured by artificial darkness. The original appearance of this picture may have been similar to the sketch in Figure 6. Olympus Mons is further to the right in this photo than on any of the other pictures in this series. The gradual progress of Olympus Mons has aided us in putting these pictures in their proper chronological order.

Marshall B. Gardner may have been the first author to state publicly that Mars photographs appeared to show the light of a central planetary sun shining through polar openings (134). Two of the eight Mars photos from his book are shown here, enlarged, in Figure 7. The larger photo that appears with them is from George Gamow's book *A Planet Called Earth*. Gamow's caption states: "The white spot at the top is the polar ice cap." Aside from the caption, I could find no mention of this photograph in the book. So why was it included in a book supposedly about Earth? I surmise that the Mars photograph was telling us something about Earth, something that Gamow himself was not at liberty to state openly.

There was no proof, when this book was published in 1963, that any ice or any water at all existed on Mars. The proper scientific protocol in this circumstance would have been for Gamow to say: "We believe the protruding polar light may be a reflection from polar ice." But to say flatly that the bright protrusion was an ice cap was unfounded and not literally possible. After all, an ice cap could not possibly extend upward for many miles as the image seems to show. In an honest description, Gamow would have added that the protruding light may have come from some unknown light source. What he actually said was not characteristic of the careful way scientists usually talk, and this may have been a deliberate sign on Gamow's part. He was quite distinguished, as one of the originators of the big bang theory.

Two protruding blobs of polar light can be seen in the Mars photo from *Life Beyond the Earth*, shown in Figure 8. Though this photo is extraordinary for its peculiar appearance, the authors gave it a mundane caption, expressing a hope that future space missions would help us to understand Mars better. One of these bright blobs of light is probably on the night side, because this polar illumination causes the perimeter of this image to be visible for well over 180 degrees around the edge, though the image is not quite in full phase.

SIGNS OF A HOLLOW VENUS?

Evaluating Venus is complicated by the perpetual cloud cover which obscures almost all the surface at any given time. In November 1980, *Life* magazine carried a full-page picture of a model of Venus under the heading: "Lifting the Veil of Venus." Numerous small glimpses of the surface captured by the Pioneer probe had been assembled to create a model of what Venus would look like without any cloud cover. At the top was a circular depression with a flat bottom (see Figure 9). About this polar basin the caption said this was "terra incognita, out of range of Pioneer's radar." The problem with this was that there was no reasonable excuse for any portion of the surface of Venus to be out of range of Pioneer's radar.

We can imagine the NASA scientists in conference as the Pioneer probe approached Venus. One or more of them would have said: "Be sure that Pioneer doesn't go into an equatorial orbit because then we would get no information on the temperate or polar regions." So they would command the probe to go into a tilted orbit that would cross the equator in addition to passing over regions above and below the equator.

What makes no sense would be to imagine the scientists saying: "Don't bother to tilt the orbit far enough to capture images of the polar regions. No polar orbit is necessary here because people don't care what the poles of Venus look like." Even if the scientists actually thought the public didn't care about the appearance of the poles, the scientists themselves should have been curious about this portion of the surface. One can't imagine them all agreeing that none of them had any interest in what Venus looked like above a certain latitude. A polar orbit would have produced images of the entire planet. Why waste taxpayers' money by doing only a percentage of the job?

The polar basin on the model of Venus looks like the beginning of a polar opening. I used this picture on the front of my first book *Hollow Earth Mysteries and the Polar Shift.*

After I sent a copy to Isaac Asimov it appears he may have given a covert response, as we shall see.

Another thought-provoking picture of Venus turned up in an immense book with many planetary photographs, the title of which I did not write down. Any large color photograph of Venus might provide the same evidence. The planet was covered in a corkscrew-shaped swirl of creamy-yellowish clouds. I wondered why the south polar clouds appeared brighter than all the rest, in spite of never receiving the sun's rays directly overhead. It seemed likely that the polar clouds were illuminated by a central planetary sun shining through a polar opening.

ARE THERE PHOTOGRAPHS SHOWING EARTH TO BE HOLLOW?

One particular photograph has captured the imagination of hollow Earth theorists and devotees. I agree that it is the most dramatic looking photographic evidence for a hollow Earth, but I don't consider it to be very strong evidence.

NASA produced a series of composite photographs of the northern hemisphere from images taken in late November 1968. Each of these was a combination of three photographs taken over a 24-hour period and fitted together to create the illusion of daylight on all sides of the planet at once. Most of these composites showed gradual shading leading to a dark center, since the pole is in shadow from November through January. This type of photograph was not strong evidence of a polar hole, though Brinsley Trench and others claimed that it was (151).

Trench was most impressed by the one photograph showing a dark center with well defined edges and a complete lack of shading leading to that center. This certainly looked like a well defined hole, but that could have been an illusion caused by unusual cloud formations ringing the polar center. If the dark center were actually to be perceived as a hole, one would have to ask why the central planetary sun was not visible through that hole. Perhaps it was covered by clouds, or perhaps it was cropped out when the images were fitted together. It may be that the illumination on this photograph extends farther north than would be possible in late November, but that discrepancy would not be evidence for the existence of a polar hole. I am not presenting these composite photographs because they are inconclusive, and have already drawn too much attention away from good photographic evidence. (See Figure 15.)

Two very tiny photographs of Earth from *Space Science and Astronomy* (145) are the only photographs of Earth that I know of that support the hollow Earth theory. Enlargements can be seen in figures 10 and 11. These are blurry because they are time-lapse photos using filters that admit only ultraviolet light. The caption for Figure 10 identifies the polar light on the night side as coming from "polar-aurora 'caps.'" The problem with this is that there is no such thing as a polar-aurora cap. Each aurora, which results from the solar wind being funneled to the magnetic poles, forms a ring around the axis, just as the magnetic poles tend to ring the axis. It is unfortunate that both Gardner and Trench accepted the idea that the aurora was the light of the Earth's internal sun shining through polar openings. If that were true, the aurora would indeed be a cap. But there have been numerous published pictures of the auroral ring at the north as well as the south pole. The night-side light in the far-UV photos of Earth, especially in Figure 10, does indeed look as though it probably is a solid cap. If this is true as stated in the caption, then the aurora cannot account for all of the polar illumination. The only other option for producing a solid "light-cap" on the night side must be an internal planetary sun, as we saw in several Mars photos from August 4, 1969. Mars, of course, cannot have polar auroras of any type because it no longer has north and south magnetic poles.

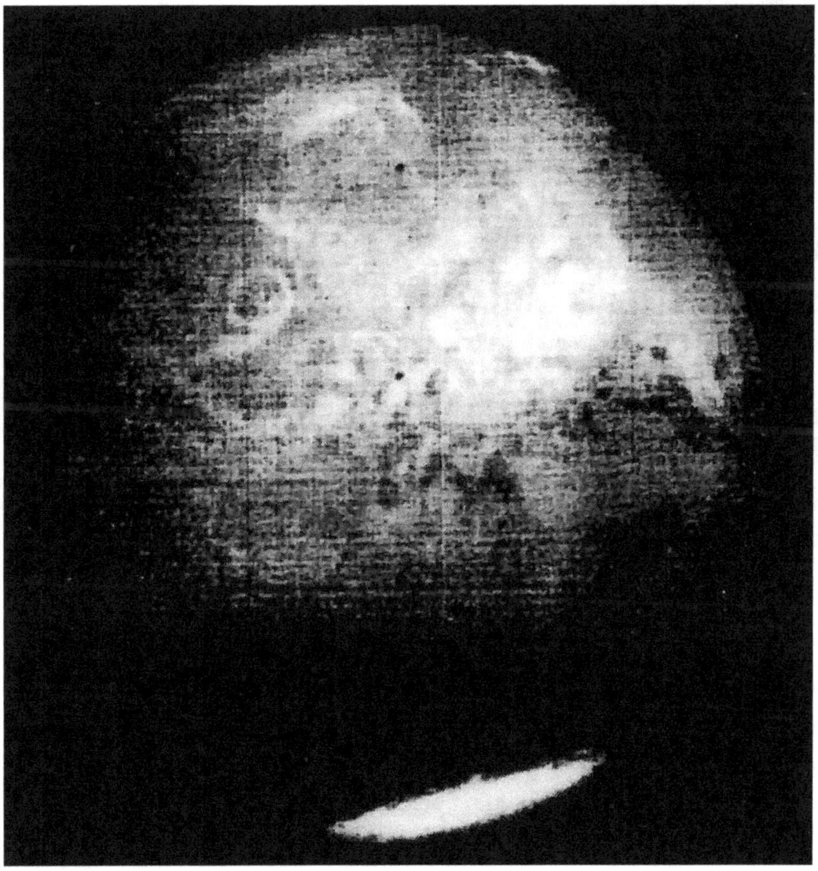

Figure 1. The Photo A, is the first in a series of four pictures taken in sequence, all showing an extraordinarily bright south polar "cap" on the night side. This picture is from *Space Science and Astronomy* (145) and carries this caption: "Mars from Mariner 7 at 293,200 mi. distance, on Aug. 4, 1969 at 10:38 UT with longitude 115 degrees centered. Nix Olympica (later renamed 'Olympus Mons') is the white circle 3/4 in. toward upper left from center. Bright south-polar cap at bottom. (JPL-NASA photo.)" The planet is in about 3/4 phase here, except that the southern hemisphere must have been artificially darkened.

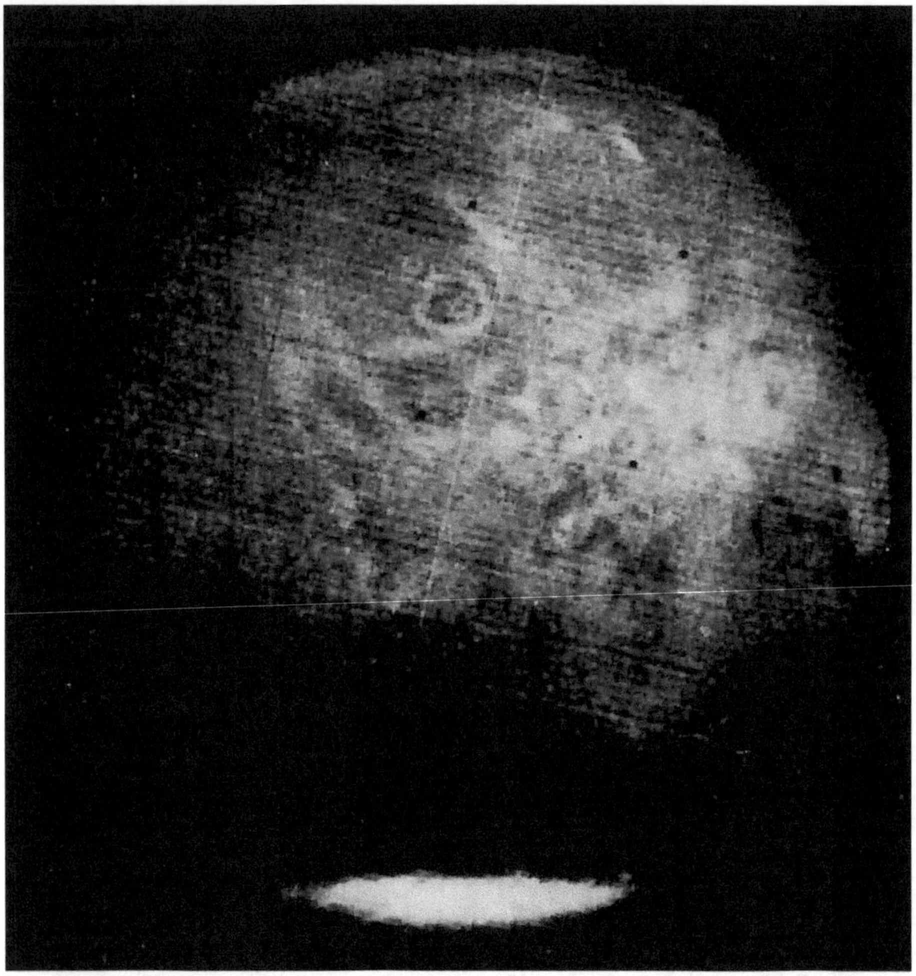

Figure 2. Referred to here as Photo B, this Mariner 7 photograph of Mars is from Patrick Moore's *Guide to Mars* (1978 edition). Both Olympus Mons and the sun's bright reflection are farther east in this picture, which is in about 2/3 phase, except for what appears to be artificial darkening of the southern hemisphere. The caption refers to "dark maria" in the southern hemisphere, perhaps in an attempt to justify the photograph's unnatural appearance. A small dot from the dot matrix should be showing near the upper right edge of the polar "cap," but I foolishly whited it out some years ago so the polar "cap" would be uniformly bright as in Photo A.

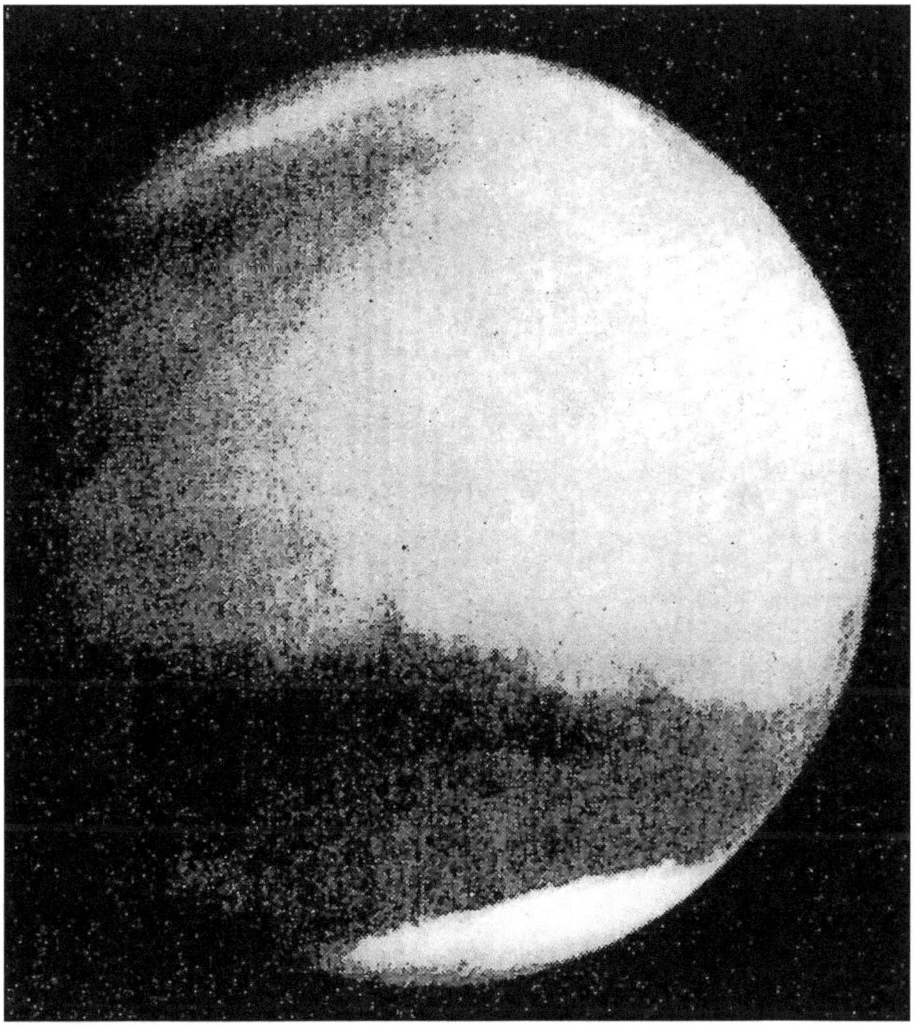

Figure 3. This Mariner 7 photograph of Mars, taken August 4, 1969, is from the *Encyclopedia Britannica*, 1974. This Photo C, as we shall call it, shows a slight progression to the right of Olympus Mons and the sun's bright reflection, as compared to their positions in Photo B. With illumination decreasing, this photo should appear darker overall than it does. It was artificially lightened, then the background sky was darkened beginning at the glitch on the right side and extending below the polar "cap" and well beyond. The polar "cap" is thus artificially placed on the day side. The day-night border was obscured by gray paint extending across the southern hemisphere. Figure 5 shows how the planet may have looked prior to retouching.

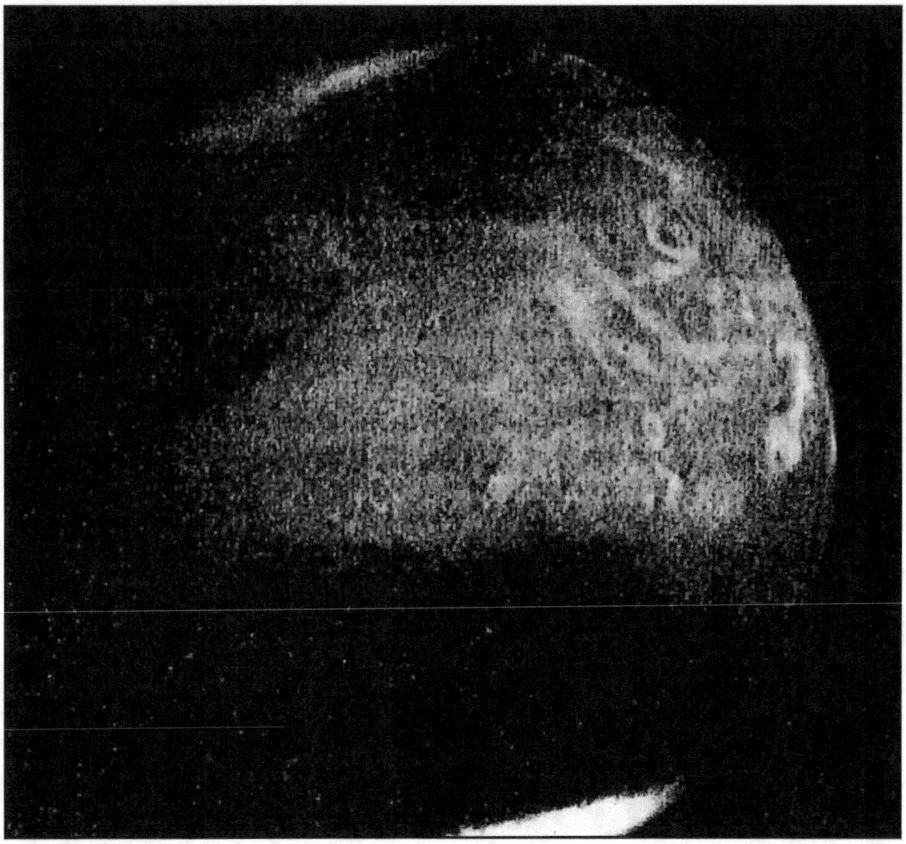

Figure 4. This Mariner 7 photograph of Mars from Aug. 4, 1969, which we will call Photo D, appears on the title page of *Pictorial Guide to the Planets* (137). As in Photo C, there is an artificial definition of the planetary outline extending under the polar "cap," that is, onto the night side. The impropriety of this may explain why most of the "cap" has been cropped off. The day-night border near the right edge of the "cap" has been obscured by artificial darkening. Though the brightest solar reflection point has disappeared from the right side, this photo has been made to appear as though brightness extends as far to the left as in Photo A. Before retouching the planet should clearly have been in less than half phase, as shown in Figure 6.

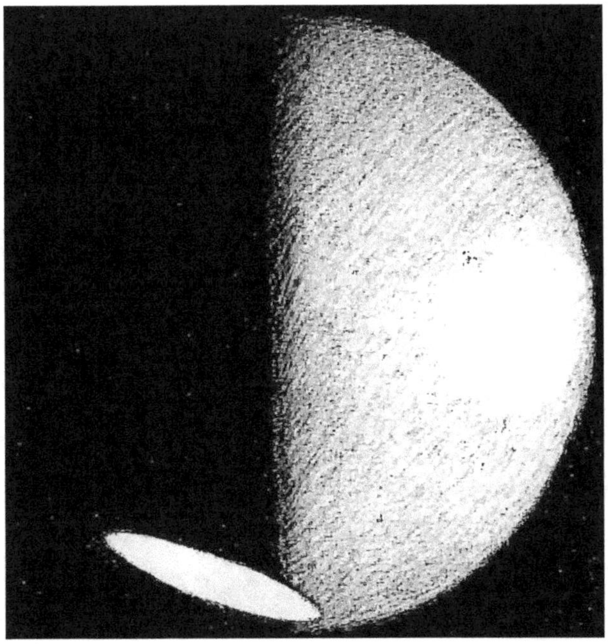

Figure 5. This sketch, based on Photo C (Figure 3) shows how that Mariner 7 photograph of Mars may have looked prior to retouching and the overall lightening of the image.

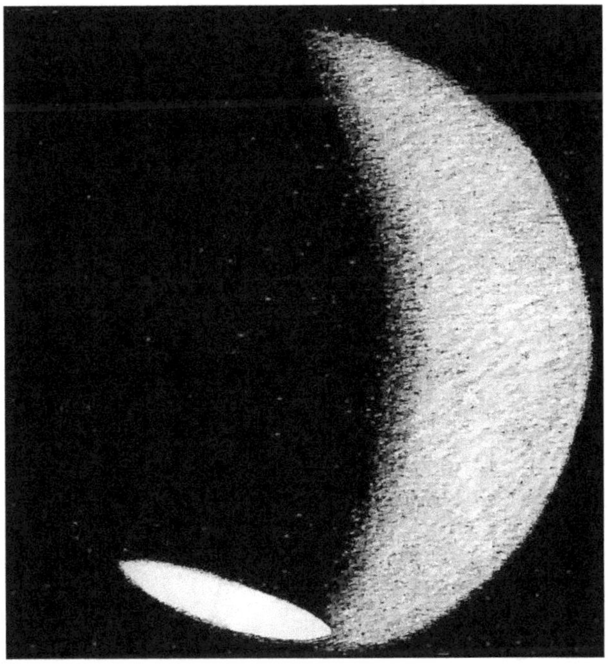

Figure 6. This sketch, based on Photo D (Figure 4), shows how that Mariner 7 photograph of Mars may have looked prior to the retouching and cropping that was done.

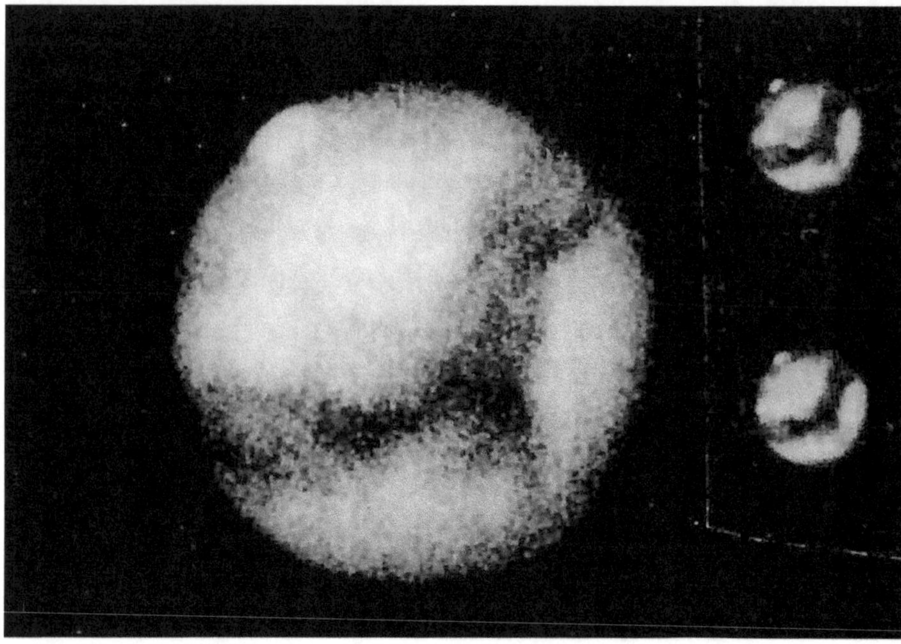

Figure 7. At left is an enlargement of a photograph taken in 1909 which appeared in George Gamow's 1963 book *A Planet Called Earth*. His caption states: "The white spot at the top is the polar ice cap." The two smaller photographs, taken in 1902, are enlargements from Marshall B. Gardner's book *A Journey to the Earth's Interior*. His caption states: "Views of Mars...showing the white circle or so-called snow-cap projected beyond the planet's surface, which precludes all possibility of its being snow or ice." This, he said, was the light of the central planetary sun shining through a polar opening.

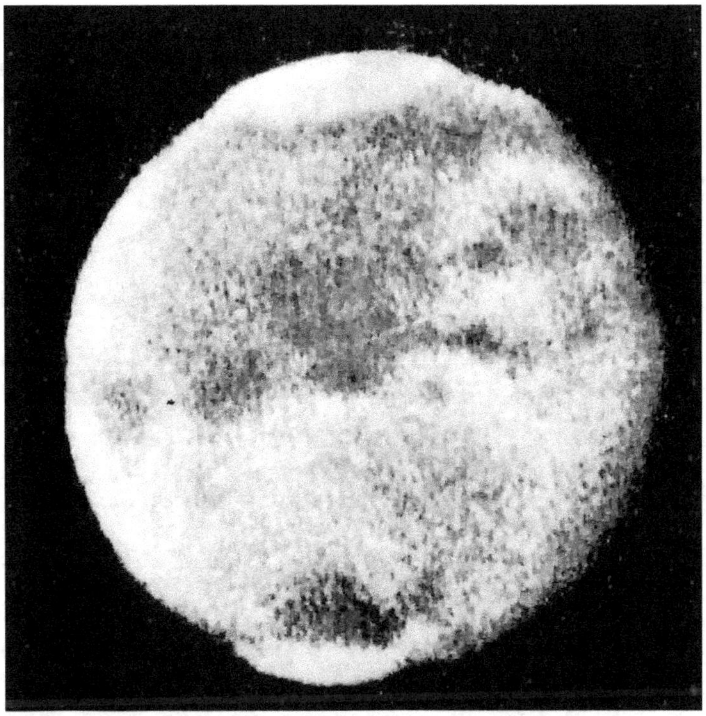

Figure 8. Two protruding blobs of polar light can be seen in this photograph of Mars from *Life Beyond the Earth* (142). The smaller blob of light seems to be off-center. If this light emanates from within the planet, it may have been partially blocked by clouds. It is difficult to explain how two polar ice caps could be visible at once, with one being off-center. The caption observes that future space missions may help us to understand Mars better.

Figure 9. This photograph of a model of Venus as it would appear without its cloud cover is from the November 1980 issue of *Life* magazine. The polar basin, which looks like the beginning of a polar hole, was said to represent "terra incognita, out of range of Pioneer's radar." I used this photograph on the cover of my 1981 book *Hollow Earth Mysteries and the Polar Shift.*

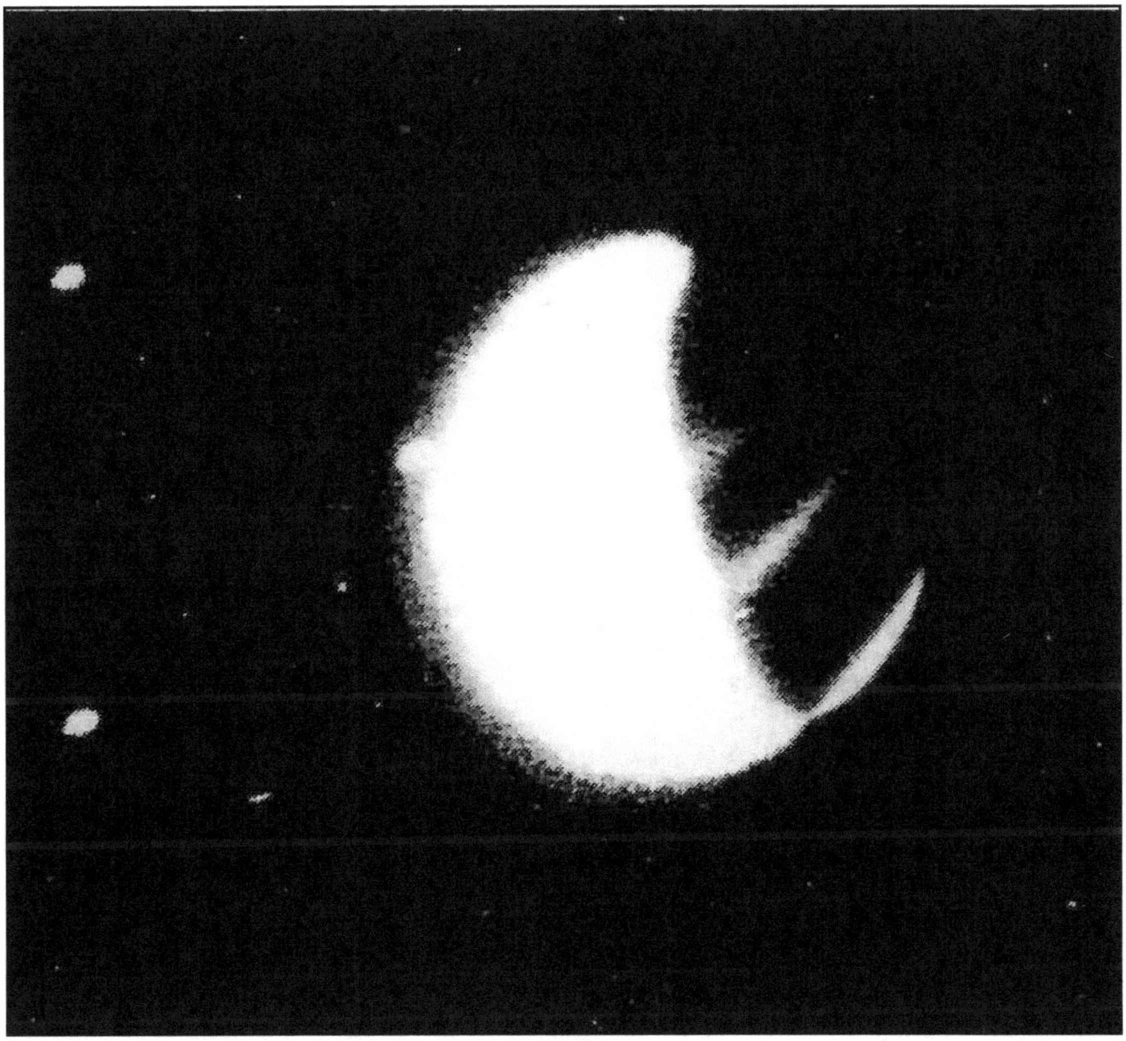

Figure 10. This is an enlargement of a "Far-UV" photograph of Earth published in *Space Science and Astronomy* (145). The pictures here and in Figure 11 are time-lapse photographs using filters that admit only ultraviolet light. The caption reads: "Fig. 133. Far-UV photo of Earth showing oxygen airglow, equatorial bands, and polar-aurora "caps," a 10-min. exposure taken from the Apollo-16 landing site on April 21, 1972, in light of wavelength 1230-1550A. The north pole is up and sunlight comes from the left. A blue star in Capricorn is very near the Earth's limb at left. (NRL-NASA photo.)"

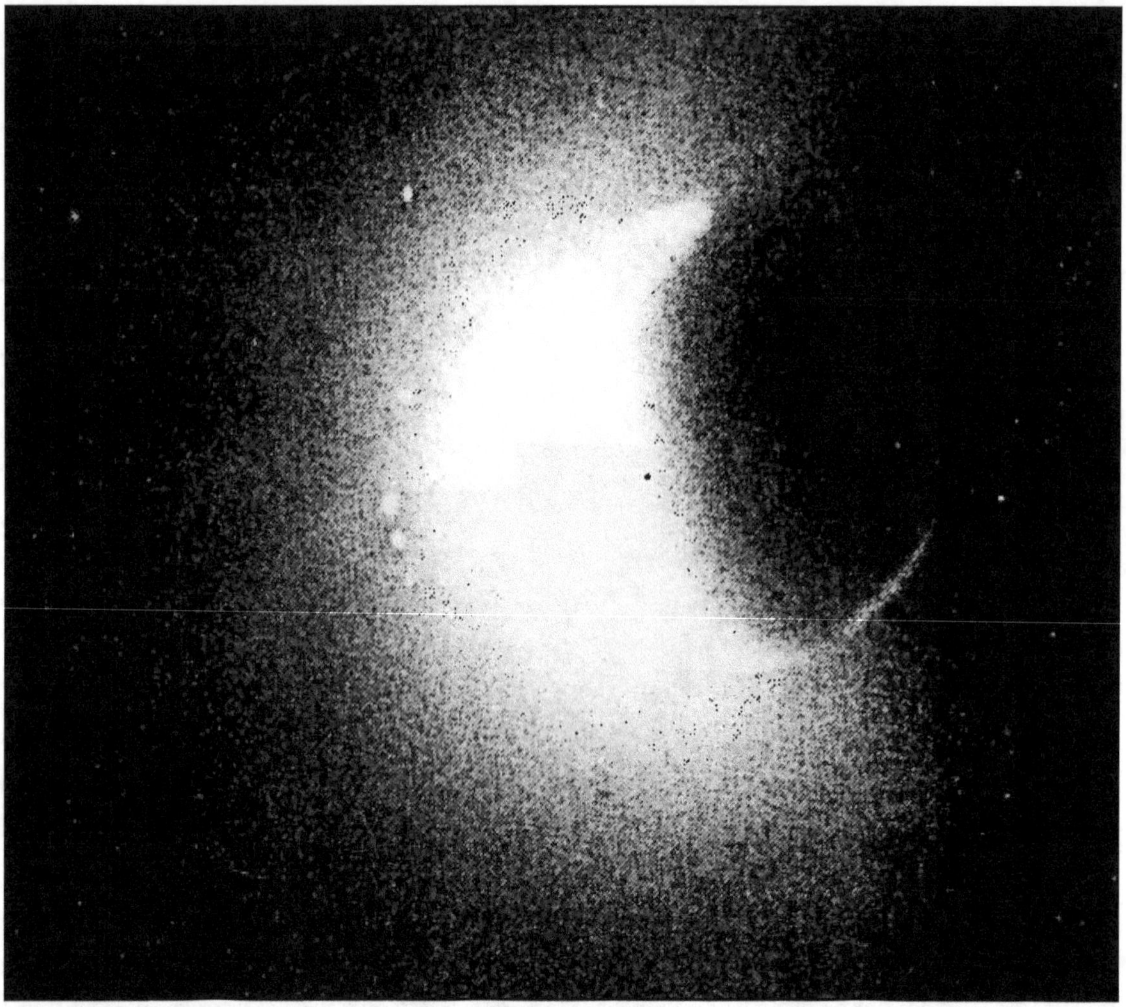

Figure 11. This enlargement of a "far UV" photograph of Earth appeared alongside the image in Figure 10, with this caption: "Fig. 134. Far-UV photo of the Earth showing the geocorona, a 15-sec exposure taken a few minutes before Figure 133 in light of wavelength 1959-1550A dominated by hydrogen Lyman-alpha (121 6A). On longer exposures the geocorona extends over 40,000 mi. (NRL-NASA photo.)" The light at the north pole here definitely looks more like a solid disc or cap than like a ring. The small original size of these two photos seemed to me an attempt to avoid getting into trouble for releasing revealing photographs of Earth showing polar light on the night side.

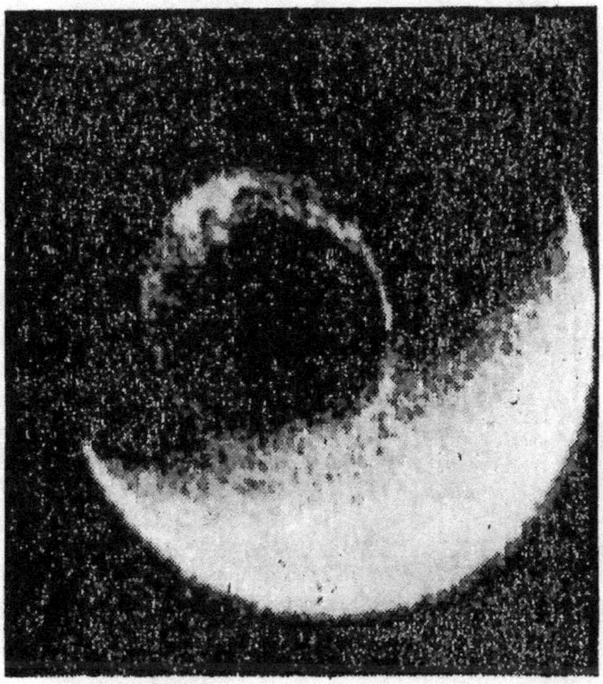

AURORA BOREALIS — This photo, taken by a NASA Dynamics Explorer satellite, shows the phenomenon known as the Northern Lights from about 14,000 miles above the Earth's North Pole. It is one of the first taken of the entire oval-shaped halo of solar particles.

Northern Lights Photos Show Polar Ring of Fire

By ROBERT LOCKE
AP Science Writer

SAN FRANCISCO — The first photographs of the entire span of the northern lights show a blazing ring of color dancing about the North Pole like a fiery halo.

* * *

"We were very excited by these images," Louis A. Frank of the University of Iowa said at a news conference during an American Geophysical Union meeting here.

The northern lights are great sheets and smears of green, red and blue lights that shimmer spectacularly across the skies of Alaska, Canada and, frequently, much of the continental United States.

At times the whole horizon seems to be ablaze, leading to suspicions by the ancients that the gods were battling across the heavens and lighting the sky with their great, flashing swords.

But earlier satellites have shown that the light show is caused when the solar wind, a stream of electrically charged particles from the sun, is funneled by the Earth's magnetic fields into the atmosphere.

The particles — electrons like those in any electric current — crash into atoms in the atmosphere and cause them to glow. The lights rise and fall in intensity according to changes in solar activity.

Frank and other members of a team from Goddard Space Flight Center in Maryland described first results of two instrument-laden Dynamics Explorer satellites.

The probes, they said, have revealed explosive storms that occur suddenly, broaden and brighten great chunks of the northern lights, then fade away within an hour.

"Hot spots" appear in the ring of light, and one surprising picture showed "a fantastic hot spot" where an intense beam of energy from the sun was exciting oxygen in the atmosphere, Frank said.

Figure 12. This polar photograph of the aurora borealis was a newspaper article published on December 10, 1981 by the Schenectady, N.Y. *Gazette*. The magnetic field's shape funnels the solar wind to the magnetic poles, where this powerful stream of charged particles sets off the auroral display. The aurora is strongest at one point (presumably the magnetic pole) but rings the axis, strongly indicating that the solar wind in fact defines the axis.

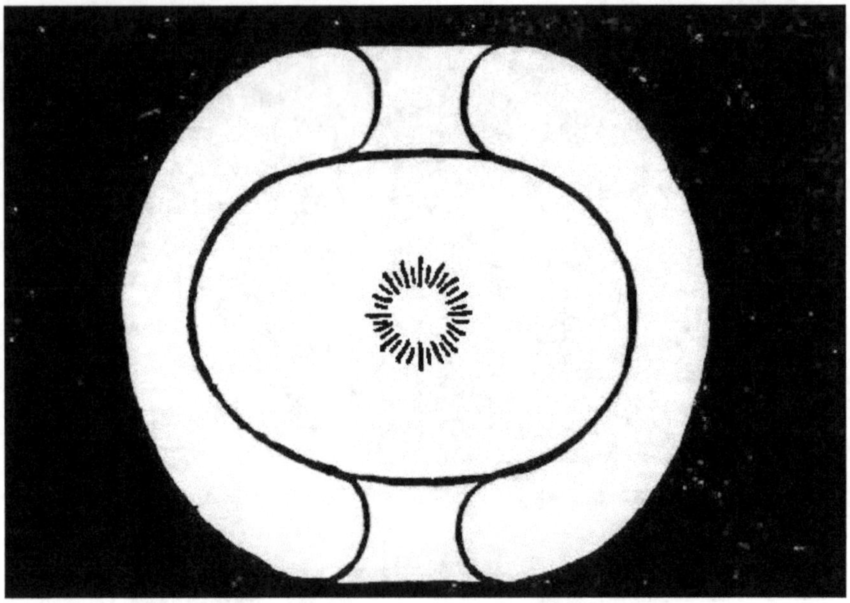

Figure 13. This is a cross-section of the structure of a spheroidal planetary nebula, which shows what the structure of Earth must have looked like in the early stages of its formation, when it was still in a nebular state. This would also be the structure of a hollow planet.

3.

WHY THE EARTH MUST BE HOLLOW, AND YET AT THE SAME TIME CANNOT POSSIBLY BE HOLLOW

Knowing what we know about the formation of the solar system from a nebula, it would be natural for us to conclude that Earth, Mars and Venus also began as miniature nebulae, with hollow shells orbiting their central stars. This is almost a no-brainer. But this idea is not placed before us.

Here is how the solid Earth theory was established as accepted truth. The National Geographic Society offered a prize for the first explorer to reach the north pole. The main competitors were Admiral Robert E. Peary and Frederick A Cook. Cook's claims from a 1908 expedition did not hold up as well as Peary's, and so Peary received the prize for an effort in March 1909. Peary later confessed that he was not certain he had reached the pole, as reported in the St. Petersburg, Florida *Times* on October 13, 1988. The compass needle gives unreliable readings in the far north and south, according to Raymond Bernard (126). The key point here is that no member of the National Geographic Society had ever set foot on either pole at the time this prize was offered. It was simply <u>assumed</u> that there was solid ground at the planetary axis.

The main arguments for a solid Earth today appear to be:

1. The weight, or mass, of the Earth is too great to allow for a huge inner hollow area.
2. Gravity would tend to fill in the Earth's interior if it were hollow, and therefore it must be solid.
3. The existence of polar holes could not be kept secret in our time, and therefore there must not be any polar holes.

These are not strong arguments. The first argument requires us to accept on faith someone's assessment as to the weight, or mass of the Earth. If scientists can be pressured into describing ice caps on Mars when there is no evidence of any ice caps, then they can be pressured into lying about Earth's weight, or mass.

The second argument ignores the possible influence of centrifugal force in keeping the planetary shell in orbit around the central planetary sun.

The third argument is purely social. It assumes that the limited number of people who have access to polar data cannot in fact be pressured to lie about it. I am not willing to make that assumption, in view of the testimony of certain explorers, and in view of the covert messages I have found from Isaac Asimov and Carl Sagan. There is also photographic evidence that should not be lightly dismissed.

Would a hollow Earth theorist like myself be permitted to publish her views for many decades if there were a cover-up as to the existence of polar holes? Most people would tend to doubt that I would be left alone to defy a cover-up of this nature, and therefore there must not be any cover-up, and the Earth must be solid.

This argument relies on too many unproven assumptions. Let us say the high king Oz is approached by his chief henchman, who thinks it is time to silence Ruth Leedy Carr. The year is 2012.

> H: How about that hollow Earther Ruth Liddy—Leedy? Oh well, she's married now anyway.
>
> Oz: Thanks for looking out for trouble, Hench, but really, the writings of this little rebel don't worry me too much. I have had many hours of innocent merriment reading her semi-scientific output.
>
> H: You said back in the '80s that you doubted she would ever discover the riddles of Isaac Asimov and Carl Sagan. Well, she has found quite a few of them.
>
> Oz: Yes, but what percentage has she deciphered correctly? Let me remind you, these messages are so weird that they all sound false even when they are correct.
>
> H: She obviously is totally untrained at codes and riddle interpretation, not to mention her lack of training as a scientist. (I did take a second major in psychology.)
>
> Oz: That's right. Her efforts are quaint and full of errors, and everyone who reads her stuff wonders what could motivate a person to make such a spectacle of herself.
>
> H: There's no question people are suspicious of anyone who is trying to save the world. Such people are considered crazy and potentially dangerous.
>
> Oz: Ms. Carr also gives people plenty of other reasons to consider her to be a kook.
>
> H: So you're not worried about this hollow Earth theorist interfering with your plans?
>
> Oz: Look at it this way, Hench. At the rate she is progressing, it might take her another fifteen years to become good at deciphering the riddles of Asimov and Sagan, and fifteen more years to convince a few people she is right, and by that time she'll be coming up on the age of 100, suffering from every ailment known to man, and besides, her cataclysmic warnings will be too late to do anyone any good. Better to leave her alone than raise questions about too many hollow Earth theorists dropping out of sight.
>
> H: You're right again, Chief, as always.

More reasons could be cited as to why the powers that be might not consider me to be a serious threat to the cover-up of a coming hollow planet disaster. One could even speculate that the ruling powers are tired of keeping us under their thumbs and don't need to dominate us anymore, and would be secretly relieved if the cover-up were ended. At any rate, any arguments on the hollow planet issue that are based on assumptions about human behavior are pretty worthless. It is better to look at the evidence with an open mind, leaving those assumptions behind.

WHY THE EARTH MUST BE HOLLOW

Scientists cannot account for the Earth's rotation. They are still arguing about whether Earth's spin has been coasting ever since the planet's condensation from a nebular state initiated that spin billions of years ago. The solar wind is the only powerful and consistent outside force that could be sustaining Earth's rotation. When that stream of charged particles becomes more powerful due to a solar storm, there is a corresponding change in Earth's rotation rate several days later, or so it was reported by John Gribbin and Stephen Plagemann in their bestseller *The Jupiter Effect*.

The solar wind is deflected by the Earth's doughnut-shaped magnetic field which funnels the charged particles to the magnetic poles, where they set off the polar auroras. These auroras don't just form at one point. They form a ring around the axis, which supports Bernard's conclusion in *The Hollow Earth*, based on magnetic field maps, that the magnetic poles themselves tend to ring the axis to some extent. The auroral rings are heavier at one point, which must correspond to the magnetic pole (see Figure 12). Therefore the solar wind's impact is greatest at that particular point in each auroral ring.

The net effect of the sun's pushing on the magnetic poles must be to boost rotation speed, though Gribbin and Plagemann do not tell us that, nor does Walter Sullivan (150) in his discussion of abrupt changes in rotation rate. There are two reasons why the solar wind must boost rotation more than interfering with it.

1. According to the principle of resonance, a force is more effective when it acts to support the motion already taking place in the target body. The same force is less effective when it opposes the target body's existing momentum.

2. The geomagnetic field forms a funnel leading to each magnetic pole. The Earth's spin causes that funnel to be slanted. That slant promotes the solar wind's access to the magnetic pole at those times when the impact could increase rotation speed. At those times when solar wind impact can decrease rotation speed the funnel is slanted away from the sun, which limits the solar wind's ability to impact the magnetic pole in a direct way.

Scientists would surely like to give the solar wind credit for sustaining the Earth's rotation. The problem is that they cannot explain what holds the axis steady as this pushing is taking place. There is no second force holding the axis steady. Without that, the solar wind should simply define an axis as close as possible to the magnetic pole, in which case the solar wind's impact would be too close to the axis to drive rotation.

The hollow Earth theory allows the solar wind to perform the double function of defining the axis and sustaining rotation.

The axis cannot conform precisely to the magnetic poles because there is no land at the axis. The polar holes create a separation between the axis and the magnetic poles. This separation leaves each magnetic pole far enough from the axis so that the solar wind's pressure at the magnetic poles can drive rotation. At the same time, the magnetic pole does to some extent form a ring that surrounds the axis. By surrounding the axis the solar wind pressure at the magnetic poles can define the axis without coinciding with it, just as we can spin an object on an axis of our choice by applying pressure with five fingers in a circle surrounding that axis.

The hollow Earth theorist does not need to supply a lot of proof that a hollow planet would receive support for its rotation. This is because the structure of a hollow planet is identical to the structure of a spheroidal planetary nebula, whose orbiting shell is "thinner at the equator than at the poles" (132). The night sky provides plenty of proof that such structures do indeed receive support.

MARS ROTATION RECEIVES NO SUPPORT ANYMORE

The planet Mars, unfortunately for us, receives no support for its rotation any longer because its dipole (north-south) magnetic field has died away. There is no overarching, doughnut-shaped magnetic field funneling the solar wind to north and south magnetic poles. Without the support the solar wind used to provide where most needed, the planet's rotation is rather rapidly slowing to the point where the outer shell will not be spinning fast enough to remain in orbit around the central planetary sun. We do not want Mars to disintegrate into gigantic asteroids and a liberated central fireball. Or at least we want to know about such a situation in advance so we can attempt to deal with it. The riddles of Isaac Asimov and Carl Sagan provide us some insight into this threat, as we shall see.

HOW THE HOLLOW EARTH THEORY EXPLAINS THE AXIS TILT

A solid planet that bulges out at the equator should not have an axis tilt. The sun's gravity should latch onto the massive equatorial region, pulling the axis permanently upright. The hollow Earth theory explains why that does not happen. The orbiting planetary shell should be thinner at the equator than at the poles, just as is the case with the orbiting shell of a spheroidal planetary nebula. The *Encyclopedia Britannica* (1974) informs us: "H.D. Curtis of the Lick Observatory found in 1918...that oblate, spheroidal or truncated shells, thinner at the equator than at the poles, accounted satisfactorily for the appearance of many planetaries."

With superior mass in the temperate regions instead of in the bulging equatorial region of a hollow Earth, the sun's gravity should attract the temperate regions, alternately, tilting the axis and giving us our seasons. If the axis were to shift at any time, the mass of the shell would be redistributed by centrifugal force, so that the axis tilt would be re-established to conform to the new normal pattern of rotation.

A HOLLOW PLANET COOLING SYSTEM FOR AN OVERHEATED EARTH

Our planet has been compared to a gigantic nuclear reactor. All seismic activity, including the reversing of Earth's magnetic field, is driven by the decay of radioactive materials in the interior. To avoid a complete meltdown of the radioactive materials, the Earthly nuclear reactor needs a cooling system. Simply allowing the reactor's walls to crack from the internal heat does not lower that heat enough to bring the reaction under control. The hollow Earth theory provides for a dramatic lowering of internal heat in a way that the solid Earth theory cannot match.

A geomagnetic field reversal is recognized as a catastrophe because there is evidence for the extinction of undersea creatures in such an event (150). The exact nature of the catastrophe has not been laid out for the public except by the anonymous geologist consulted for Jess Stearn's book on Edgar Cayce. This expert said he had been pushing colleagues unsuccessfully to recognize a link between the reversing of Earth's field and the shifting of the axis, also popularly known as a polar shift.

After a prolonged period of decay in geomagnetic field strength—2000 years or more—the location of the magnetic poles wanders erratically until the north magnetic pole settles somewhere in what was previously the southern hemisphere and the south magnetic pole settles in the former northern hemisphere. It should go without saying that the axis would be relocated to conform to the new magnetic polar locations as part of this process. It should go without saying because it is so obvious that the axis is now defined by the solar wind striking in a circle (the aurora) surrounding the axis. The axis can't be any closer to the magnetic poles, where the solar wind strikes, because there is no land at the axis. As Marshall Gardner put it, orbiting matter cannot remain stationary at the axis.

All this is elementary. The cooling system for the Earthly nuclear reactor consists mainly of the migration of polar openings to conform to the new location of the magnetic poles. The openings must migrate as a result of centrifugal force acting around the new axis of rotation. The resulting disturbance in the crust would release a lot of internal heat, and the flow of ocean water into the cracks that are created would result in further massive cooling of the interior. This cooling must take place to enable the geomagnetic field to regain its normal strength, since heat destroys magnets. It is the increasing tempo of radioactive heat production in the interior that has led to over 2000 years magnetic field decay (see Bloxam and Gubbins). The geomagnetic field does not just spontaneously regain its strength as a result of a reversal. That reversal must first bring about a dramatic cooling of the interior. Another factor producing a dramatic internal heat release would be the repositioning of the equatorial bulge to conform to the new axis. This change must explain why Edgar Cayce's source predicted widespread eruption of volcanoes in the "Torrid area" to occur along with the formation of a new axis. Several signs of an approaching geomagnetic field reversal are described in detail in the December 1989 article by Jeremy Bloxam and David Gubbins (128). Core spots (also called hot spots), possessing reversed flux features periodically emerge near Indonesia and follow a track toward the southern Atlantic Ocean, where they are "growing in size and strength and moving southward." The article timidly observes that Gubbins believes "this may be the process that eventually reverses the Earth's field." This spectacle of hot spots with reversed magnetism gathering in the southern Atlantic Ocean turned up in the 2005 novel *Polar Shift* by Clive Cussler and Paul Kemprecos. The novel did not portray a literal shift in Earth's axis, which disappointed me.

There can be no doubt that Cayce was predicting a genuine shift in the planetary axis. This is shown by the climate reversals described in this prediction:

"There will be upheavals in the Arctic and in the Antarctic that will make for the eruption of volcanoes in the Torrid area; and there will be then the shifting of the poles so that where there has been a frigid or semi-tropical climate there will be a more tropical one, and moss and fern will grow." When asked what he would predict for the year 2000 his answer was a polar shift.

We can speculate that the polar upheavals described by Cayce are the disturbances that will occur as polar openings begin shifting to conform to changes in the location of the axis. The fact that the prophecies do not spell this out, and do not spell out the source of the light that is predicted to appear in the clouds after 1998, which coincides with the time of the axis shift, shows that there was and is powerful secrecy at work relating to the hollow Earth evidence. The anonymous geologist (149) had no comment to make either on the predicted cloud-light or on the prediction of polar upheavals to coincide with an axis change. Evidently, he knew better than to touch something as sensitive as the hollow Earth theory.

4.

SOS SIGNS IN A "HOLLOW EARTH" ESSAY

"...WHEN IT RATTLED ACROSS A HOLLOW WORLD..." – ISAAC ASIMOV

It was no surprise that Isaac Asimov denounced the hollow Earth theory in his essay entitled "The Hollow Earth" (64). I received advance notice, in early 1986, that he was going to write on the subject.

I was fairly well known among readers of fringe literature in the early 1980s as a result of writing three books and having them sold by Robert Girard of Arcturus Books. Also the first two were republished by Gray Barker of Saucerian Press, who had published *The Hollow Earth* by Raymond Bernard. I communicated with editors of UFO literature from time to time. My third book bore the subtitle *Asimov's Warning* after the main title *Hollow Earth Apocalypse*.

A young man called me from New York and said Asimov was going to reply to my work in a magazine essay. I didn't expect to ever find that essay, and yet eventually, in a library, I picked up Asimov's essay collection *Past, Present and Future*, and there it was.

It was somewhat unbelievable that some associate of Asimov's would be aware of his intention to publish an essay referring to me and would take the trouble to find out my real name (my pseudonym was Floria Benton), track down my telephone number and notify me at some expense to himself that Asimov was going to reply to my work. After a short discussion he said he had to go somewhere on his bicycle. Another associate of Asimov approached me in a print shop in Chestertown, Maryland in 1987, as we shall see.

I was fairly sure there would be some kind of secret sign endorsing the hollow Earth theory in Asimov's "Hollow Earth" essay. I had already found broad hints from Asimov. In a rambling description of a thunderstorm in his 1983 novel (114), he had written that its noise "rattled across a hollow world." He had to know that that kind of language would be seized on by hollow planet enthusiasts like me. I had sent him my 1981 book. Whether he received by 1983 book before finishing his 1983 novel, I cannot be sure. But I had praised him in 1981 for his courage in dropping hollow Earth hints in defiance of a cover-up. If he really considered the hollow Earth theory to be one of those unscientific and ridiculous notions that he was so dedicated to opposing, he should never, never have used the term "hollow world," even if only as a metaphor.

CIVILIZATION'S END BY 2010— ASIMOV'S PREDICTION

Asimov had predicted the imminent end of civilization in a 1980 interview published in *Mother Earth News*. This was where I first noticed his hollow Earth hints. On the front of the September- October issue was this headline: "Thirty Years to Armageddon?" He did predict civilization's demise by 2010, but elsewhere in the interview he said it was only civilization as we know it that was going to end by that time.

Space travelers of the future will build settlements within asteroids, he said, adding that these pioneers "will hollow out the mini-planets." This was the biggest hollow planet hint. He also said "we are used to living on the outside of the hull" of Earth if it is viewed as a spaceship. Future space pioneers will be "living <u>inside</u> a world," he said, speaking of their enclosed existence. Living inside a world is not the same thing as living in an inside world, but of course it is possible that he was misquoted. At any rate, the hollow planet language combined with Asimov's prediction of civilization's end convinced me that he was aware of a coming hollow planet disaster that would bring civilization to its knees. The photos published with the interview showed him looking somber.

I had been studying the cataclysmic prophecies of Edgar Cayce in relation to the hollow Earth theory. Now all of the sudden it seemed that Asimov was hinting that this scenario was correct. It seemed he was flaunting his knowledge without being willing to share it openly with the public. Then it came to me that the opposite was true. He was desperately trying to share his knowledge, to warn us, but hinting was all he could do. I burst into tears and wept for a long time. I began scouring his work and that of Carl Sagan for hints which I reported again and again in books, and later in newsletters.

The hollow Earth theory was bashed by Asimov in his "Hollow Earth" essay, as one would expect. The final two sentences were as follows:

"So the hollow Earth theory is a potpourri of poopery and an inanity of insanity. So say I!" The last sentence, containing only six letters, had begun with S, O and S. Could this not be an SOS sign indicating his criticisms of the hollow Earth theory had been made under duress, and he really supported the theory? After reporting that he had been accused of dropping hollow Earth hints (referring to my 1985 book), why would he end the essay with something that looked like a big fat hint negating all his criticisms of the theory, unless it really was a hint? The exclamation point, if rotated, consisted of a dot and a dash, the basic elements of Morse Code from which the SOS sign is derived. That final sentence can be converted to "'SOS,' say I!" with the addition of only one letter. The invented word "crackpottery" could be a cataclysmic hint, since it calls to mind a hollow earthenware pot that is cracking.

Taking a closer look at "potpourri of poopery" and "inanity of insanity," these expressions are modified ABA forms in which the third part is almost identical to the first part. The SOS sign is a simple ABA form where the first and third parts are identical. The letter that changes the "popery" sound to poopery" is an "O," whereas the letter that changes "inanity" to "insanity" is an "S." So these ABA-type expressions feature the letters of the SOS sign as the pivotal letters.

Turning to the beginning of the essay, there are two words containing S, O and S in that order in the sentence describing Asimov's reaction to receiving my book with his name in the title. He said that just when he thought he could no longer be surprised and nothing could break life's round of dullness, "something astonishing comes along and socks me right in the funny bone." Both "astonishing" and "socks" contain S, O and S in that order. If you think that was just a coincidence, try looking through his work for another sentence with two such words in it. (In his poem "The Author's Ordeal" we find a rare use of "so's" as well as the use of the word "socks.")

The Morse code "S" consists of three dots, and we know that three dots can indicate a pause in written works—a pause or perhaps a sentence that is left unfinished. Asimov's story "Old-fashioned" (95) contains numerous sets of three dots, with a period or question mark added. The story itself involves the hero sending SOS signs from the asteroid belt to Earth, requesting rescue because a spaceship is disabled. The sets of dots seemed to come in pairs, amounting to two "S" signs in Morse code, separated by a line or so of text in which a word suggesting the letter "O" could be found. These words were: hole, orbit, oh, and pill. I could find "SOS" given three times this way, using a total of six sets of three dots, grouped together in pairs.

It seemed clear to me, on studying the story, that Asimov was sending an SOS to the people of Earth, just as his hero was sending SOS signs to Earth. As to the exact message, the story portrays danger to Earth from a small black hole that could potentially grow and threaten Earth if not dealt with. This could not literally be what Asimov was warning Earth about because small black holes have not been proven to exist, and such a thing would probably have been undetectable.

The dangerous object was portrayed as being located in the asteroid belt, and so Asimov could have been indicating that an asteroid-like object was a threat to Earth. Since a black hole is made of compressed stars, Asimov could have been indicating that a star or star-like object was a threat to Earth or would become so. I thought he could have been referring to a future threat from the collapse of Mars into asteroids and a central star that would be free to roam in our direction. This is a possibility, but it is not confirmed by the character names that Asimov has provided. It is not possible to spell "Mars" from the names "Ben Estes" and "Harvey Funarelli," but it is possible to spell "Venus."

COULD MARS AND VENUS BOTH POSE A THREAT TO EARTH?

In my study of Asimov's riddles, I have found several that can spell both Mars and Venus. These are: "Levular Mandamus" (79), "Minerva Schlump" (16), and "Suranoviremblastiran" (34), and then there is the sentence "Horsabelum desoderatim andevidual stinko" from "The Critic on the Hearth" (55). It so happens that one of these possible riddles does involve a threat to Earth. Mandamus sets in motion the gradual destruction of Earth as a habitable world in *Robots and Empire*.

In his name we can find: "Venus's small sun and Mars's small sun are near us." In "Minerva Schlump" we can find: "Venus an' Mars have small inner suns, are limp, can crumple." The letters for "inner" are given in order. In "Suranoviremblastiran" we can find: "Venus an' Mars are liable to turn into suns an' meteors." In "Horsabelum desoderatim andeviduali stinko" we can find: "Venus and Mars are both liable to krumble into asteroids and suns and threaten Earth in time."

Long riddles can have numerous solutions, but at least we can say Asimov's riddles provide some support for believing both Mars and Venus are potential threats to Earth. In "Ben Estes, Harvey Funarelli" we can find:

"In fifty tu ninety years Venus's inner sun is liable tu threaten Earth." Once again, a message like this may be intended, or perhaps not. This message does fit well with the plot of the story "Old-fashioned," as we have seen. Estes and Funarelli discover a small black hole in the asteroid belt which not only damages their spacecraft but also could grow and pose a danger to Earth if not diverted onto a new path. The reason Estes sends an SOS to Earth is not only that he and Funarelli need to be rescued, but also that Earth needs to be warned of this danger, about which Earth knows nothing.

Both Isaac Asimov and Carl Sagan presented riddle names containing the syllable "Sos." Since very few names contain this syllable, it is possible that SOS signs are intended here. In "Charles Soskind" (21) from Asimov we can find: "Each riddle I do is a riski SOS sine." Soskind's fellow character "Marcus Quintus Trent" may convey this message: "Scientists are quiet as tu a sun in Mars. It is a secret."

Carl Sagan's novel *Contact* contains the names of ethnic and religious groups concerned about a possible doomsday, including "the Sossafer Chasids," which may contain this message:

"SOS. The Earth's acsis is to shift after for to ate decades. (This catastrofe cood destroi societi.)"

We can find numerous sets of three dots in Sagan's novel as the heroine speaks on the phone to her colleague Ian Broderick about a signal from space. Is there any sign of a Morse code "O," or any "O" in the intervals between these dots? Yes there is, and Sagan does better than that. In the interval where three hyphens are found, representing the letter "O" in Morse code, Sagan has the heroine refer to dots and dashes. This definitely reminds us of Morse code and reinforces the SOS sign he has created using two sets of triple dots separated by a set of triple hyphens, representing dashes.

The name "Ian Broderick" may shed some light on Sagan's reason for sending an SOS sign. The name can spell "broken brain." In the books I sent to Sagan I noted that his book title *Broca's Brain* sounds like "broke his brain," This seemed to me to be a clear signal that scientists were under some kind of pressure to keep secrets, backed up with a threat that their brains would be broken if they defied the cover-up. In his book Sagan had focused on the brain of an anatomist named Paul Broca, whose brain had been preserved in a jar. The name "Ian Broderick" may contain this message: "Brain cood be broken and I cood die." If the name of Broderick's countryman "Annunziata" is added to his name, forming a longer riddle, the message may be: "I cannot be direct. I need to uze a code. Brain izn't broken, but it can be."

The heroine Eleanor Arroway is also involved in this conversation where Sagan portrays an SOS sign. In her name the message may be: "Yoo learn really well, R.A.L." My name then was Ruth Anne Leedy. Adding the name of Arroway's dad "Theodore F." (Arroway) to the riddle, the message may be: "Yoo learned of the hollow Earth theory, Rooth Anne Leedy, and yoo learned really well." An alternate solution for "Eleanor Arroway" would be:

“Eye warn yew, allow no one near Norway.”

Several of Asimov’s calls for help did not involve an SOS sign as such. In “Brinsley Sheridan Cooper” (107) we can see: “Brain is disabled. I need yoor help. Please decipher and share riddles.” In “Nellie Griswold” (62) we see: “End o’ world riddles need reeding. No one is willing.” In “Nephi Morler” (79) we find: “Help me help people hoo R in peril.”

Asimov’s favorite hero “Hari Seldon,” from the Foundation novels, has the letters for “riddles” grouped together in his name. In this name we can find: “Hard riddles or easi ones, all are SOS sines. I need aid.” It has been my honor to be able to attempt to help him out.

5.
HOW ISAAC ASIMOV ENDORSED MY HOLLOW PLANET INTERPRETATION OF KING ARTHUR'S GRAIL LEGENDS

There is a lot of mystery surrounding the end-time prophecies of Christ and around grail traditions in general. Was a real cataclysm being predicted, or some kind of a spiritual event? The truth always seems to be out there, just out of reach.

I believe Isaac Asimov confirmed my hollow planet interpretation of Christ as a solar figure surrounded by twelve zodiacal followers. I described this interpretation in my first book, which was sent to Asimov in 1981. I wrote as follows:

"In *Mysteries of the Holy Grail*, Corinne Heline states that King Arthur, surrounded by the twelve knights of the Round Table, is said to represent symbolically the sun surrounded by the twelve signs of the zodiac. Christ, surrounded by the twelve apostles, would be in the same symbolic position. How do we know that the sun in this symbolic description is inside the Earth? It is quite simple. The signs of the zodiac exist only from the perspective of the Earth. Seen from other points in the universe, the constellations would be arranged differently. Even though the changes might be slight at short distances from the Earth, it is clear that symbolically any object, such as a sun, placed at the center of the signs of the zodiac must be represented as an object on—or inside—the Earth."

In his novel *Foundation's Edge* (1982) Asimov included a table of twelve. Christ was known to have sat at a table with his twelve disciples, one of whom departed leaving twelve figures at the table. A character named Stor Gendibel was one of the figures at Asimov's table of twelve. "Stor" is an archaic form of the word "star," suggesting that the figures at Asimov's table are all stellar, or zodiacal, figures.

In my next two books I did a diagram of the table of twelve described by Corinne Heline, whose work I had discussed in 1981. Heline placed a chalice at the center of King Arthur's Round Table, said to be a recreation of Christ's Last Supper table. The absence of one of the knights left twelve figures at the table, counting Arthur. The knights were said by Heline to be zodiacal figures, with Arthur representing the sun. Inside the central chalice was the blood of Christ, which on Good Friday glowed with the radiance of a sun.

For my 1983 and 1985 books I sketched Heline's version of Arthur's Round Table, with twelve zodiacal stars surrounding it and a circle in the middle representing the rim of the chalice, and the sun-like blood inside that circle. I commented that this sketch looked like a polar view of a hollow Earth surrounded by zodiac signs, with a central planetary sun visible through a polar opening.

Asimov recreated this entire image in his 1986 novel *Foundation and Earth*, using an actual planet. This was his most powerful confirmation of my interpretation of grail symbolism.

In this novel, a spaceship called the Far Star hovers over the southern pole of the planet Comporellon. From that position six bright entry stations resembling stars can be seen on the night side, spaced out, with six more presumably spread out in the day side, so that a complete circle of twelve entry stations are arranged in a ring around the planet. One entry station leaves its position and approaches the Far Star, thus echoing the "empty seat" motif described by Heline and also found with Asimov's table of twelve in 1982, and of course also found at Christ's Last Supper.

By recreating my grail table diagram using a planet, with star-like entry stations and a central "star," Asimov was putting himself on record as confirming that I had correctly interpreted the hollow planet symbolism created by Heline and by Christ himself. We may note that Heline explicitly said Arthur's table represented the Earth, and she endorsed the hollow Earth concept by saying a certain hollow mountain had a channel leading to the center of the Earth.

A JUNIOR SUN SHOOTING LIGHT AT THE NORTH STAR?

Various hollow Earth and axis shift symbols can be found in Christ's teaching and in his life story. 1 want to mention another grail legend which demonstrates that Christ was not the only one creating symbolism along this line. A sketch of an Egyptian sun god shooting at a north polar constellation can be found in Flavia Anderson's book *The Ancient Secret: In Search of the Holy Grail.* (Note how she defined the grail as a secret, rather than simply a mystery or a physical object.) Ra, the son of the sun god, was shooting arrows representing light beams at a polar constellation known in Egypt as the Thigh or the Haunch of the Bull. She went on to say this constellation must be the source of numerous grail legends about solar heroes receiving thigh wounds. Be that as it may, the real meaning of the sketch seemed to me to be that a junior "son" of the big sun sends light beams to a polar constellation because these are the only stars he can see from his vantage point inside the Earth. In other words, the sketch places the junior sun in the interior of a hollow Earth looking through a polar opening.

The Hebrew Jacob, according to Anderson, must have been considered a solar hero because he had twelve sons, who corresponded neatly to the months of the year (or the signs of the zodiac). The implication seemed to be that Christ too was casting himself as a solar figure by choosing twelve disciples.

Heline too must have considered Christ to be a solar figure, since King Arthur with his twelve knights of the Round Table were a recreation of Christ's Last Supper in her view, and she specified that Arthur represented the sun.

CHRIST'S RETURN IN THE CLOUDS—A SOLAR PHENOMENON?

If Christ deliberately cast himself as a solar figure by choosing twelve disciples, then his promise to return in the clouds during a cataclysm was probably not meant to be perceived as a promise that he would be seen in the clouds in human form. It is more likely that he was telling us, in a roundabout way, that the sun he represents will appear in the clouds. Since his zodiacal followers define him as an Earthly sun (the constellations are arranged differently if not seen from the perspective of Earth), then he must have been saying the inner Earth's sun will shine through polar openings and be reflected back to ground level by the clouds. Marshall Gardner described the polar "water sky" as reflecting the inner sun's light.

How do we know Christ was describing a shift in polar openings resulting from an Earth axis shift? We can presume that the inner sun always shines into the polar clouds to some degree. But that doesn't mean it is visible to large numbers of people. Very few people travel to the coldest regions on Earth, and people do not settle there. Christ was describing a sudden appearance that many people would be able to perceive. In other words, a polar hole would be relocated to a populated area and its light would be seen in the clouds by the people living nearby.

Hollow Earth symbolism is associated with Christ from the time of his birth. According to some authorities he was born in a cave that was used as a stable. A solar figure located in a hollow within the Earth is an image that resembles a hollow planet to some extent. The virgin birth created hollow Earth imagery, since a virgin was a commonly used symbol for a planet in ancient times. A solar figure within the womb of a virgin-planet is an image resembling a hollow planet. Christ told people they are the "light of the world," then commented that one does not light a candle and place it under a bushel, but on a stand. In this metaphor the candle is compared to the light of the world and the world is compared, theoretically, with the hollow bushel basket containing a central fire.

Christ's cross was a symbol of the world axis, according to mythologist Joseph Campbell (129). This stunning revelation appears in fine print in a footnote—just another example of people being skittish about revealing hidden truths. If Christ viewed his cross in this way, his crucifixion can be looked on as a dramatic demonstration that a sun, represented by himself, is united with the world axis, represented by the cross. Campbell also noted that the "world mountain" represented the world axis in ancient times. With this in mind we can speculate that Christ was creating axis shift imagery when he said that faith can move a mountain.

Christ's end-time parable of the wise and foolish virgins can be seen as a portrait of a solar system in which all virgin-planets shine like lamps, but not all are lighted from within. By including ten virgins in the parable Christ appears to have been showing that he was aware of the existence of more planets than most people in ancient times knew about. Considering that the asteroid belt was possibly a planet at one time, or would have been without the gravitational pull of Jupiter, and considering that another planet may exist beyond Pluto, Christ's number was about right. If he meant to be telling us that some virgin-planets have internal illumination, and if he knew how many planets existed altogether, the parable shows that his cataclysmic prophecies were based on authentic advanced knowledge, not just a rehashing of the wild guesses of other prophets.

In the Book of Revelation, Chapter 12, a dragon's tail is portrayed sweeping stars from the heavens. This brings up the whole subject of Earth axis shifting, as we shall see in a moment. The dragon is menacing a woman clothed with the sun, with the Moon beneath her feet and a crown of twelve (zodiacal) stars on her head. She must represent Earth because of the zodiacal stars and because she has the Moon in a subordinate position. A hollow Earth would be clothed internally with sunshine, and half clothed with sunshine externally.

Helen P. Blavatsky notes in *The Secret Doctrine* that a constellation known as the Dragon has been described as sweeping stars from heaven with its tail during Earth axis shifts or "pralayas." All that really means is that the constellation's "tail" assumes a new position when the heavenly bodies change courses, and so the entire change has been blamed on the dragon's tail. The real reason for all changes in the heavenly bodies is that Earth has become oriented differently in space due to an axis shift. Blavatsky's insight that the dragon represents a constellation whose tail is said to sweep stars from heaven tells us that the dragon referred to in the biblical passage mentioned earlier must indeed represent the Dragon constellation. And so the passage is really describing an Earth axis shift.

Star shifts are explained in *The Impending Golden Age* by Rev. John Lowell and Rev. Kaspar Nureddin of the Sanctilean Syntonium (a branch of freemasonry). They write:

"Today our North Pole points approximately at a star called the North Star. Following the tilting of the axis of the Earth, our North Pole will no longer point at this star. The star will not have moved. The axis of the Earth will have tilted to cause the star to appear to have moved." Also they write:

"This tilting is popularly called the Star Shift. The shifting of the axis of our Earth results in what appears to be the shifting of all the stars in the heavens." In this process, of course, some stars would drop out of sight permanently, while others not previously visible in one's hemisphere would begin to be seen.

The shifting of Earth's axis is accompanied by the supervision of humanity by higher beings, according to Lowell and Nureddin. Blavatsky also writes of this, as follows (p. 372):

"According to Plato, in order to obtain clear and precise ideas on royalty, its origin and power, one has to look back to the first principles of history and tradition. Great changes, he says, have occurred in days of old, <u>in heaven and on Earth</u>, and the present state of things is one of the results (Karma). Our traditions tell us of many marvels, of changes that have taken place in the course of the sun, of Saturn's reign, and of a thousand other matters that remain scattered about in human memory, <u>but one never hears anything of the EVIL</u> which has produced those revolutions, nor of the evil which directly followed them. Yet...that Evil is the principle one has to talk about, to be able to treat of royalty and the origin of power."

Both Blavatsky and Brinsley LePoer Trench included hollow planet material as well as axis shift material in their books. I didn't see where either of them drew the obvious inference that an axis shift on a hollow Earth would involve a shift in polar openings due to centrifugal force. Maybe the thought was too horrible for them to entertain it. Or perhaps this secret was too dangerous to be spoken about.

6.
"RUTH A. LEEDY IS A PROPHET." "YOU COULD LEARN TO DECIPHER A RIDDLE. I CAN BE YOUR TEACHER."

"Tear the Earth apart." These words can be found in "Theptar," the month when civilization was predicted to end in *Nightfall*, by Isaac Asimov and Robert Silverberg. This is a short puzzle, actually part of a longer puzzle. This one made-up word is so packed with warning signs that are so easy to see that we need to spend some time on "Theptar" all by itself.

Asimov was a science writer in addition to writing fiction for his entire career. His story "Nightfall," on which the novel was based, was one of his early stories, published while he was still in his twenties. Young as he was, he was portraying the end of civilization on a fictional planet, but putting the name of Earth into the month when that disaster was foretold to occur, and when it did occur. On Kalgash the disaster was a total eclipse plunging the planet into chaos. Not only can Theptar spell "Earth," but it can also describe a violent physical change on Earth. "Tear the Earth apart." The entire riddle is "the nineteenth of Theptar," which is the exact date when Kalgash falls into chaos as predicted. This riddle appears to be telling us: "Noo pattern of rotation to tear the Earth apart near the open portion in the north." This message is confirmed by similar messages from the same novel. What gives these messages such power is the fact the end of civilization has been so closely tied to these riddles, and the fact that this was done by a highly competent and respected science writer. This particular riddle not only concerns the end of civilization, but actually involves the <u>prediction</u> of that end, which is exactly what we find when we decode the message. We find Asimov himself <u>predicting</u> an event that is likely to throw civilization into chaos.

In order to use the exact date of civilization's fall as a riddle, Asimov had to imagine an event that could actually be defined as happening on one particular day. This is why he had to portray an eclipse instead of a geological catastrophe. The main point, however, is that this riddle does indeed carry a prediction for the fall of civilization. Through easy and obvious riddles, which we will be looking at, Asimov shows us that he was a master riddler. So none of this was an accident.

"Shift" and "Earth" and "rotation" can be found in "Siferra, Theremon," the names of two characters who fall in love during the crisis on Kalgash and find hope for the future. A long message can be found here:

"Earth is to shift soon in its rotation. I am not free to share that information. It is safer for Rooth Anne to share it for me. Mars is soon to shatter into meteors. Its inner son mite then hit the Earth. I fear this threat the most."

The prophet who announces the date for Kalgash's fall is named "Folimun," alias "Mondior." In these two names we can find: "I inform U of doom in mi riddl. If U no find mi riddl, million mor of U R doomd." Folimun leads a group called "the Apostles of Flame." Here we can find: "Help me to help masses of people to sta safe." In "Altinol" and his former employer "Morthaine" we find: "Northern hole on hollo Earth to harm mani in time." Three characters who become uncivilized during the crisis on Kalgash are "Butella, Fridnor, Talpin." The message here may be: "Upper portion of Europe about due to fall into open pole." A longer solution to this would be: "Upper portion of Europe be about to fall into an open pole. No land to be in upper Europe after nu rotation pattern on our planet." In "Veloran, Sten" from the same novel, we find: "One learns never to reveal a reversal's near."

RUTH A. LEEDY—A HOLLOW EARTH POLAR SHIFT PROPHET?

The letters for "prophet" do not turn up often in riddle names. To find two riddles close together in the same story that can spell "prophet" is unusual. In his final story in *Magic* (54), Asimov strongly suggests the phrase "is a prophet" in the name of the fairy Misaprop. She gives a clumsy blessing to Prince Delightful. In "Misaprop, Delightful" we can find: "Ruth A. Leedi is a prophet. I am grateful that she has desiphered mi riddle messages as to a polar hole shift that is to shatter upper Europe." Prince Delightful befriends a dragon who agrees to go to battle against the Lotharingians, who are then defeated in the town of Papeete. In "Lotharingians, Papeete" we find: "Rooth Anne L. is a prophet. She is right on the hollo Earth theori, also on a polar shipht that is going to happen soon.

Asimov died in 1992 without knowing whether I would ever discover his riddles. He hoped I would do so because I was so dedicated to publicizing his cataclysmic hints. In one of his last stories (52) in the name "Cuthbert Cantrip Culloden;" he seems to encourage me to study the riddles, with his help: "Ruth Anne Leedi, U could learn to decipher a riddle. I can be ur teacher."

This is a good time to show how both Isaac Asimov and Carl Sagan sought to draw us all into an understanding of their riddles, by offering us some easy examples. As our foremost science educators, they both had a gift for making complex material understandable and fascinating.

Two short animal names in Asimov's final novel *Forward the Foundation* are obviously riddles. We realize "lamec" must be an anagram of "camel" when we are told a lamec is a beast of burden. We realize "greti" must be an anagram of "tiger" when we read that riding a greti is dangerous because if you fall off it might eat you. In "Manowell," from the same novel, published after Asimov's death from cancer, we can find: "No am well." Manowell is said to have been a bloody emperor. From "Sam Aickman," in a novel published the year Asimov died (122), we find this: "I am a sick man." In "Red, Slim;" two short names from "Youth" (112), Asimov seems to be telling us: "Reed mi riddles." Two characters who are lost in space in *Forward the Foundation* may have gone to Voreg or Sarip. Sarip is obviously an anagram of Paris. In "Voreg, Sarip" we can find: "I grieve over Paris." This is confirmed by the grief resulting from the two characters being lost in space, plus the death of another family member. This message indicates that Paris is far enough north to be devastated by the formation of a new polar opening in the upper portion of Europe.

The name of Asimov's wife Janet Jeppson was clearly used to create the name "Tejan Popjens Lih," as we have seen. We also saw a scrambled version of "Azimov" inserted into the name of a wise planetary leader (34). It is surprising to find "Isaak Asimov" in the brief name "Hamish Sankov" (111). The complete message appears to be: "I, Isaak Asimov, hav a hi mission." With "mission" in Sankov's job title of commissioner of Mars, we can be sure that this message is correct. "Lors Avakim" (31) is another short name where we find "Isaak Asimov." The message may be: "Asasil rools Isaak Asimov." Asimov wrote numerous stories featuring a powerful demon named Azazel, as we shall see. There are many riddles indicating that Asimov was under the influence of an outside force. In "Lavinia Demachek" (114) we find: "I am invaded. I am like a machine." Demachek's predecessor "Albert Minnim" carries a similar message:

"Mi brain be bent, malleable." Similarly, in "Sindra Lambid" (79) we find:"Mi brain is disabld."

Asimov's book title *I, Robot* seems to be relevant here. In "Hubert Mandel" (98) we can find: "Tu rule the Earth U need tu learn tu bend and tear the human brane." Likewise, "Tomas Bistervan" (79) may carry this message: "Asimov's brain is in tatters." Also, in "Sortibackenstrete" (84), in a story where a character is to be "dismembered" for science, we can find: "Scientists' brains are bent, torn and broken." In "Fortescue Quackenbrane Flubb" (58) we see: "Our ruler can cause quakes, catastrofes. Our branes are not free, can be bent or broken." Flubb's brain is adjusted by the demon Azazel, resulting in a painful dream leading to his death. In "Charlie Norbert" (29) we see: "I cannot tell anione that the Earth be hollo. Brain be torn." His partner "Craig Levinson" carries a related message: "Everione in science is living a lie."

Asimov's career ambition seems to be expressed in a message from "Mortimer W. Jacobson" (100), where we find: "I want to become a major science writer." Asimov seems to be telling something about himself and his wife in "John Semper Travis" (101), as follows: "Asimov moves into Janet Jeppson's home, then marries her." He did move in with her after separating from his first wife. In "Joseph Olsen" (62) he seems to tell about Janet again: "J. Jeppson helps hopeless people." Janet Jeppson was Asimov's psychiatrist.

Carl Sagan describes his career as an astronomer (he specialized in studying Venus) in "Peter Valerian" (148), where we find: "I peer at everi planet." Valerian is, in fact, an astronomer. We have seen how Sagan's prisoner name "Tyrone Free" can spell "Eye R not free." Tyrone has a visitor named "Abonneba Eda," and if his name is added to the riddle the message may be: "Eye be not free. Brane be torn and abraded." The unusual word "abraded" is used less than 50 pages after Tyrone and Abonneba are mentioned. In the names of two friends, "David Drumlin, Helga Bork," Sagan may have been saying: "R.A. Leedi, I am glad U are OK, and no have been murdered or had ur brain broken." Similarly, Asimov observes in "W. Bradford Hume" (74): "We feared U'd be harmed or murdered, R.A." In "Tessa Anita Wendel" (63) Asimov's message appears to be: "Leedi is well, wasn't assassinated."

7.
"I, ASIMOV, AM NOT VERY GOOD AT SAVING THIS EARTH." "RUTH HAS A HARD JOB TO DO, HAS TO DO A GOOD JOB."

"Ewe hold a hollow world in ewer hand, Rooth Anne Leedi." This message can be found in "Hari Seldon, Wanda" (30). Similarly, in "Rutilan Horder" (114) we find: "U hold a hollo Earth in ur hand, Ruth Anne Leedi." Also, in "Athor, Nyilda" 120) we see: "Yoo hold a hollo Irth in yoor hand, Rooth Ann L." In "Myron Dynast, Ginny, Harold" (72) we find: "Yoo hold a hollo Irth in yoor hands, Rooth Ann L. Many a tim I told yoo this."

Could Isaac Asimov really have believed that I held the fate of the world in my hands just because I sent him a book or two or three on the hollow Earth theory in relation to Earth axis shifting? Surely it wasn't just that. I was crucial to the future of the world because I made some important points without getting killed in the process. The chapter ahead of this one ended with this message: "Leedi is well, wasn't assassinated." Asimov and Carl Sagan were evidently holding their breath after receiving my books, hoping the ax would not fall on me.

Asimov himself was trying to save the world from widespread death and destruction. In "Namarath Godhisavatta" (34) he puts his name on a vital message: "I, Asimov, am not vari good at saving this Irth." Rooth Ann might sav it. Godhisavatta's partner and subordinate "Jogoroth Sobhaddartha" appears to convey this message:

"Rooth has a hard job to do, has to do a good job." These two messages set forth clearly what Asimov was trying to accomplish in his riddles, and what role he was hoping I would play. It was a hard job because I had to do it alone without a lot of clear guidance or financial support, and any mistakes I made would tend to discredit my entire project.

"Ur mission is risqué business." This is from "Morris U. Bunque" (52). In "Quindor Shandess" (34) we can see: "R.A. has risqued as no one has e'er done."

"RUTH ANNE LEEDY" IN CARL SAGAN'S RIDDLES

We have seen a solution to "Stefan Alexeivich Baruda," the doomsday alarmist in Carl Sagan's novel, with my name not included in the message. Now here is an alternate solution: "Ruth Anne Leedi, U are rite. A field reversal shall cause Earth's axis tu be shifted bi the sun in the near future." In "Eleanor Arroway, Theodore F." we can find: "Rooth Anne Leedy, yoo are free to defend the hollow Earth theory. Eye are not free." In the same novel we saw "Eye R not free" in the prisoner name Tyrone Free. Arroway and her dad were paired in the preceding riddle, while her boyfriend is paired with her in the next one. In "Eleanor Arroway, Kenneth der Heer" we find: "Rooth Anne Leedy, yoo are korrekt on the hollow Earth theory."

It appears Isaac Asimov praised my boldness, as follows: "U are truthful, a fighter, a real leader, Ruth Anne Leedi," This is from "Delightful, Laurelene" (54). Similarly, in "Faraday, Laurelene" we see: "R.A. Leedy, U are unafrade. U are really free."

Shortly after my first book was sent to Asimov in April 1981, the name "Eustace Bartholomew Wasservogel" appeared in a story, of his entitled "Can You Prove It?" (22). The message appears to be: "We got ur booc. Because U can see that Mars must be hollow, U seem to believe that Earth must be hollow as well." This message goes along with the story title, which is asking for proof. It is helpful that the name can spell "hollow Earth," with both words spelled correctly. A version of the word "hollow" is visible in the name Bartholomew.

Asimov's mystery "Hot or Cold" (118) also appeared shortly after I sent him my first book. In the names "Simon Brooke, Lucas J. Atterbury" we can find: "R.A. Callanan, you can make books on cataclysms. It is not easy job. You are smart." He seems to have noticed that I had glued the book together myself with "rubber cement." The word "book" is strongly indicated in the name "Brooke."

My safety was apparently of great concern to Asimov. In "Dulcinea Greenwich Glitz" (51) we find: "Ruth Anne Leedi, U are in real danger. A little accident with a gun is easi to arrange." In "Edgar Andrev, Ernett Second," (79), where the first character is almost shot by the second, the message may be: "Secret dangers are ever near. Don't test gods," In "Newton G. Descartes" (57) we find: "Ewe saw secret dangers, R.A. Ewe were not scared." I actually have been afraid many times. In "Giskard Reventlov" (114) we find: "Danger to R.A. Leedi is konstant and veri grave." An alternate solution for "Giskard Reventlov" would be:

"R.A. Leedi is risking all to serve as savior."

In spite of the hazards I faced in speaking out on the hollow Earth and cataclysm, Asimov must have thought I was safer doing so than he would have been. In "Sunmaster Fourteen" (66) he explained what a risky position he was in: "Am not as safe as U are, R.A. U are not famous, so U are freer." Since he was famous and his word was trusted, he was watched more closely and was not permitted to step out of line.

A BEGGAR AND A BAG LADY—LEEDY FALLS ON HARD TIMES

I was out of work almost totally from 1986 to the fall of 1992, due to several things that had stressed me out. I had been writing books and gluing them together, playing the organ for a busy Catholic parish, moving several times, and dealing with a stubborn TMJ jaw problem. I was hoping to bounce back, but was settling into poverty, living on government benefits, and didn't seem to be making progress. I began dating my future husband John Carr in 1989.

Asimov's riddles reflected his concern over these developments, I believe. In "graben" (120), an animal about to be eaten by a desperate refugee, this message may be present: "R.A. be a brane. R.A. be a beggar." In "Charles F. Cabbage" (52) we can find: "As a beggar R.A. be safe. As a Carr she'll be less safe." This may mean that as a beggar I was less of a threat to the cataclysmic cover-up and therefore relatively safe.

"Leedy, ye be a bag lady. Maybe I gambled badly." This is from "Mabel, Bill, Gladys" (43). When Asimov did riddles addressed to me, he was gambling on my ability to share their message with others. If I fell apart and failed him, those riddles might not be helpful to any other interpreter of his work who didn't know about me. The riddles he did generally involved writing a work of fiction. If he included riddles aimed at me that were unlikely ever to be understood, then he had made a poor investment of his time—a bad gamble.

"Be strong, R.A. No be a beggar." This is from "Rogen Benastra" (66). In "Yussif Newberry" (58) we see: "Yu were fine. Yu were in business." I previously had piano students in addition to being a church organist. In one riddle Asimov seems to give me a pass for being unemployed. In "Herman Gelb, Peter Jonsbeck" (40) we see: "Rooth Anne L. can make books on a hollo Earth catastrophe, be a prophet, a messenger. She has no other job." I was an obscure author, as we see in "Endor Levanian" (66): "Nearli no one ever learned ov R.A. Leedi." Asimov didn't want me to get discouraged and set aside my cataclysmic project. In "Harla Branno, Terminus" (34) we can find: "U must not terminate ur labor on the hollo Earth theori, Ruth Anne L. U are a real brain." In "Skystrip Two" (66) we see:"I worry yoo stop yoor work. Try to stik to skript." In "Sibelius Hopkins" (37) it appears Asimov wishes me well: "I hope U'll be OK. I hope U'll be in business soon," Hopkins has two partners. In "Ralph Colodny" we find: "Ay pray yoo can hold on, R.A.L." In "Ben Devers" I am guessing at this message: "Reeders be ever needed." My finances were a concern for Asimov. In "Tamwile Elar, Cinda Monay" (30) we see: "I want to aid yoo, R.A. Leedy, yet I cannot will yoo any o' my money."

In Asimov's final stories, a farewell to me can be found. "U R nice tu me, ne'er rude, ne'er a critic, Ruth Leedi." This is from "Hilderic, Ermentrude" (56). In "Marcus, Ernebtrude" (54) we can find: "U are ne'er rude tu me, ne'er curse me." In "Meliversa, Tertius" (56) we see: "U are servile tu me, treat me as a master. U treat me rite. I am sari tu leave U." In "Yugo Amaryl, Dahl" (66) Asimov may have been saying: "Yoo ar a holy lady, R.A.L. May yoo marry a good guy." In "Jonas Willard, Meg Cathcart" (41) we find: "I cannot wait to see whether John Wane Carr and Rooth Anne Leedi will get married." John and I met in 1989 and were married in 1995. In "Tryma Acarnio" (30) Asimov may have been saying: "Yoo can marry Carr, R.A. Try it."

In *Nightfall*, there are two characters, one fat and one thin, who are always together. In "Faro, Yimot" we find: "If yoo ar to marry you ar far too fat, R.A." In "Hano Lindor" (66) the message may be: "Hold on, R.A.L. Don' di."

HINTS ABOUT TWO MORE HOLLOW EARTH THEORISTS

The name Marshall Gardner lends itself to word plays. Isaac Asimov and Carl Sagan had a field day with this name.

"Martial gardeners" are created by Asimov in his final novel (30). Spies armed with military or martial weapons are able to pass as gardeners assembling to work on the emperor's estate. The gardeners have been summoned—brought together from far and wide—to serve in this way, and therefore they are "marshalled gardeners." They are arranged in order and by rank for a welcoming ceremony, which once again makes them marshalled gardeners. One of the gardeners is to officiate at the ceremony, which means he is to serve as a marshal. One of the real gardeners grabs a weapon and commits an assassination, making him more of a martial gardener than the others.

In Sagan's novel a professor assembles his students at his home and gets them to do weeding and gardening. He has thus "marshalled the gardeners." The heroine and her friend stroll through gardens with patriotic colors and statues expressing military themes. These are "martial gardens." Flowers are arranged by rank on the steps of a ziggurat, simulating the Hanging Gardens of Babylon. These are "marshalled gardens."

The "martial gardener" incident in Asimov's novel (30) involves three characters, all of whose names can spell the last names of two hollow Earth theorists, and two of which can also spell "Leedi." These theorists are Raymond Bernard and Marshall B. Gardner. In "Gleb Andorin" we find: "Bernard and Gardner be bold and braini." In "Gambol Deen Namarti" we can find: "Bernard and Gardner be bold, brite men." In "Mandell Gruber" the message may be: "M.B. Gardner be labelled a madman. R. Bernard be murdered." I believe Marshall Gardner did become a mental patient at some point after writing his best known book. Raymond Bernard was not heard from for long after writing *The Hollow Earth*. I cannot say he met with foul play, though that message can be found in the riddle. The fact that Gruber does go mad and is executed after committing an assassination tends to confirm this riddle message.

THREE HOLLOW EARTH THEORISTS IN ONE RIDDLE

The name Bernard is given to a dragon in Asimov's final fantasy story in *Magic* (54). Since Prince Delightful clings to his neck, we can assume the names "Bernard, Delightful" go together in a riddle. The names Bernard and Gardner can both be found in these letters, in addition to "Ruth A. Leedi." The message appears to be: "Bernard, Gardner and Ruth Anne Leedi are truthful regarding the Earth. I had to lie."

The name Bernard can be found in "Baldur Anderson" (8), where the message may be: "U and Bernard are our leaders, R.A.L."

We have seen riddles where Asimov appears to recognize me as a prophet. One more could be added here. In "Raindrop Forty-three" (66) he appears to say: "Rooth Anne, yoo are a prophet for the entire Earth if yoo don't die."

Few people who have seen my work would deny that it is prophetic. But to have Isaac Asimov say so makes all the difference.

8.
"MICHAEL CALLINAN MISSES YOU, RUTH." "C.T. TINCHER IS SINNER. I'M LISTENING. I HEAR HIM TELL R.L. THIS."

I was saying the things Isaac Asimov and other experts couldn't safely say about a coming hollow planet catastrophe. I was not only describing the coming dangers but also highlighting the cataclysmic hints that Asimov and Carl Sagan were providing. I placed Asimov's name in the title of my third, fourth, and fifth books.

If my efforts were to have any hope of convincing the public, Asimov knew he would have to endorse me as someone who was speaking for him. How could he do this in a subtle manner that would not result in trouble for him or his publishers? Aside from placing my name in his riddles or perhaps creating a fictional character here and there who was based on me, or perhaps on my parents, his other option was to focus attention on my first husband and other men friends.

Michael Callinan, to whom I was married for a little over five years ending in 1982, was a singer who had an interest in occult and paranormal phenomena. When I produced my first hollow Earth book in 1981, Michael began to have thoughts about writing a book of his own. We collaborated on a book called *Serpents of Fire: The Hitler - Hollow Earth Connection.* This was published by Gray Barker of Saucerian Press, who had published my 1981 book.

Our pseudonyms were Michael Alexander and Floria Benton. Working on this project with Michael really emphasized for me what I already knew about Michael's writing ability. His spelling was unbelievably poor, and so were his grammar and punctuation.

In his story "Cal" (38) Asimov presents a robot character named Cal who aspires to be a writer. His first effort produces only gibberish, then he gets a little better as a result of an adjustment. He writes a story called "The Introoder," where the spelling is barely understandable, and the grammar and punctuation are also terrible. This was rather similar to what I might have expected in Michael's writing. (This is not to say Michael lacked intelligence. He was certainly outstanding as a singer.)

It may be just a coincidence, but Michael actually could have written a story with that title. (It is likely that Asimov found out about that fact, I think, as will be seen.) Michael had been staying with me for perhaps two months when one night at 11 P.M. we heard the front door open. I froze in terror. He went out to see who it was. It was the husband of a friend of mine who had visited me once before for an evening cup of tea. This visit was to take the place of the conversation we were missing because he showed up at the Deer Park (watering hole) in Newark, Delaware just as I was leaving. This meeting at my place was a social event, no hanky panky. Nevertheless, he seemed to think he had an opening to come again, with no advance notice, and without knocking.

I knew Michael would view this incident in the worst possible light. He would think I had given this friend the green light to visit me at any time, and it would look like an affair had been going on. I immediately began yelling at my intruder friend at the top of my voice and ordering him to leave. Michael did later give me a hard time over this incident.

We have seen the riddle message I found in "Sotayn Quintesetz ABT" (34), which was this: "It is not easy to quiz anyone about you." Assuming this was correct and was aimed at me, it establishes that Asimov would have been likely to send someone to befriend Michael and get him talking about his five years of marriage with me. This dramatic incident would have been likely to come up if Michael had been asked: "Did you ever have any indication that Ruth had been unfaithful, or that she had been promiscuous?"

It is pretty obvious that someone with a last name like Callinan would be likely to take on the nickname Cal. So Asimov's use of this name is a possible sign pointing to Michael's last name. It is not a strong sign, but it can prompt us to take a close look at the name Cal uses for the hero of another of his stories. After writing "The Introoder" he is adjusted again and his writing improves. He writes "The Shiny Quarter," featuring the detective "Calumet Smithson."

DID ASIMOV ACCIDENTALLY PROVIDE LETTERS FOR "MICHAEL CALLINAN"?

The name "Michael Callinan" can be made using the letters from the name Calumet Smithson. The complete message appears to be: "Michael Callinan misses U. U must miss him too." Calumet Smithson has a long conversation with a character named "Mr. Wassell." Adding this name to "Calumet Smithson," the message becomes: "Michael Callinan misses U, Ruth Anne L. U must miss him as well." The similarity of these two messages tends to confirm both of them as correct. Taking the name "Cal" and his pseudonym "Euphrosyne Durando," we have seen the message: "A polar hole shall end upper Europe, as U and Ed Cayce prophesy, R.A.L." In the second part of "The Shiny Quarter," in the names "Mr. Foster, Lionell, Mann, Ostrak" we can find: "Rooler Asasel, master of all, is not a normal man, is a tall monkie." By itself this message is not very significant because it is so long that other messages are possible. Turning to the story title "The Shiny Quarter," the message may be: "The Sasquatsh are in the inner Earth in quantity." In other words, only a small percentage of the Sasquatch live on the outer surface. The main point from this message is that "Sasquatsh" can be found in these letters, and a Sasquatch can be described as a tall monkey.

Getting back to Michael Callinan, in "Mian Endelecki" (30) we can find: "Mickie Callinan needed Alice Callinan." Michael's mother Alice moved to Florida after retirement. Several years later Michael moved to Florida to live with her. She usually called him "Mickey." In "Claws 'Pop' McFang" (17) we can find: "Long ago M. Callanan's pop was a scofflaw." Michael's father Pat was something of a con man. At his death the ID cards for numerous identities were found on him, making it hard for authorities to notify Alice of her ex-husband's death immediately.

The robot Cal writes a third story after another adjustment. The title "Perfectly Formal" could spell this message, referring to Michael's singing: "Cal perform(s) perfectly." Referring to my piano accompaniment for Michael, the message might be: "Yoo try to perform perfectly for Mycal." This message could be expanded to refer to my work interpreting Asimov's hints and riddles. The message could be: "Yoo perform perfectly for Mycal. Try to perform perfectly for me."

A couple of riddles about my father's poetry also help to identify me as someone whose work is endorsed by Isaac Asimov and Carl Sagan. In "James P. Joule" (57) Asimov appears to say: "I am jealous o' James's poems." My father did books full of poetry, mostly adaptations of limericks. In Sagan's novel, "Palmer Joss" provides a similar message: "I am jealos o' James's poems. I am also jealos o' R.A.L.'s prose." He was jealous of the fact that I could say things he was forbidden to say.

DID ASIMOV MEAN "KALIININ" TO SOUND LIKE "CALLINAN"?

The name Kaliinin, sounding like Callinan, turns up in Asimov's 1987 novel (29). In "Sophia Kaliinin" along with "Yuri Konev," her boyfriend's name, I was able to find: "In your opinion, R.A. Kallinan, Venus is hollo. Kan you prove your kase?" This message must refer to the first book and letter I sent to Asimov in 1981, when I was still married to Michael. The book bore my pseudonym of Floria Benton, which Michael had suggested, but I included a letter signed Ruth Callinan. The cover of the book carried a large picture of a model of Venus with a polar basin that looked like the beginning of a polar opening (see Figure 9). The caption of the picture in *Life* magazine (139) had claimed the polar basin was "terra incognita, out of range of Pioneer's radar." I considered this a poor excuse for not capturing images of the polar regions, and I figured the real reason this area was not pictured was that there is no land at the poles.

Asimov's first impression of my work would have been that Ruth Callinan thought Venus was hollow, and his riddle of 1987 seemed to be recapturing that moment. As to my proof that Venus is hollow, the comments I have already made about photos of Venus are the main proof I can offer. Carl Sagan, who was an expert on Venus, portrayed Venus being called a "hellhole" in his novel *Contact*. That comment can be seen as a hint that Venus is hollow. Without that, we might conclude that Sagan thought Mars and Earth were hollow but Venus was not. The testimony of an expert like Sagan on this question is better than most other kinds of evidence that a non-expert like myself could produce about a cloud-covered planet like Venus.

One could argue that the evidence that Mars and Earth are hollow can be used to indicate that Venus must also be hollow because of its similar size. That is a credible argument, which can strengthen the case, but it does not stand up very well by itself.

If Isaac Asimov would secretly ask me, through a riddle, to try to prove that Venus is hollow, the very fact that he would make that request is evidence that Venus is likely to be hollow. If the whole idea were ridiculous, he wouldn't have asked me to try to prove the case. He was, after all, an expert and surely knew the answer.

HOW ISAAC ASIMOV GATHERED INFORMATION ABOUT ME

There was an incident in 1987 that may help to show that Isaac Asimov was trying to find out more about this author who was attempting to publicize his cataclysmic hints. I was in a print shop in Chestertown, Maryland when another customer apparently noticed that I was going to print something about Asimov. He asked me about my project, and I said I thought Asimov was hinting about a cataclysm on a hollow Earth. He replied: "I know Isaac Asimov. I am in a club with him." I must have asked him what Asimov was like. His answer was that Asimov was friendly and agreeable as long as he was the center of attention, otherwise not. The conversation went on until I loaned him my 1985 book *Hollow Earth Apocalypse: Asimov's Warning*. He returned it with comments that were neither favorable nor unfavorable.

I later reflected that it was unlikely that any fellow customer would ever ask me about my work in a print shop. It would be even more unlikely that a member of a small New York club that Isaac Asimov belonged to would turn up by chance in Chestertown, Maryland and start questioning me. This man was probably sent by Isaac Asimov as a sign to me that he took my work seriously. It can also serve as a sign to anyone who reads my account of this incident and accepts that it is a true account, not manufactured by me to make myself look important.

Sending an agent to approach me or approach my ex-husband was one way to gather information about me, but Asimov went further than that. Quizzing people about me was not easy for him, as we have seen. In order to gather more information, he could have put listening devices or even cameras in my home. I now have proof that he did that. His purpose had to be to pick up information that he could put into his riddles to help me prove that he was interested in my work. Anything he said about me would serve as an endorsement of my cataclysmic research, <u>if I could prove</u> that he was referring to me personally.

"I am listening." This possible message turned up in "Lestim Gianni" (34) as I attempted to solve this riddle around 2018. I was not sure this was correct, even though each letter was used exactly once. The message didn't seem useful because it didn't say who or what Asimov might be listening to. I couldn't help noticing "lissen" in "Blissenobiarella" from the same novel, but once again I did not see a reference to myself or the subjects I was writing about. I could see "C.T. Tincher" in "Chetter Hummin" (66), which was a possible reference to my landlord and boyfriend of 1987 and 1988, but I did not see a specific message about him. Now, however, I have found a much more definite message on all this, which has led to other discoveries. In *Nightfall*, the psychologist Sheerin comments on trauma suffered by Gistin and Chimmilit.

In "Sheerin, Gistin, Chimmilit" we can find: "C.T. Tincher is sinner. I'm listening. I heer him tell

R.L. this." The syllable "Tinch" is given here, and the words "heer" and "him." This message must be correct because one of Clay Tincher's favorite sayings was "I am a sinner," which he said with deliberate emphasis on each word. He was unmarried, had picked up women in dance halls and had served as a sperm donor, through sexual activity. His mother had arranged his first sexual experience with a "friend" when he was ten, which may have set him on a lifelong course of sexual adventure.

Finding this message caused me to take another look at the Lestim Gianni riddle, which now seemed to be a definite message from Asimov that he was listening to me. Gianni is one of two minor characters who sit at the table of twelve leaders of the Second Foundation (34). The other minor character is "Leonis Cheng." In his name we can find: "1 lissen. I see eech one hoo goes in." Another character at the table is "Stor Gendibal," whose name offers many possible messages. On the subject of listening, we can find this: "I be listening to R.A. Leedi."

HIGH PRAISE TO ME FROM ISAAC ASIMOV

Gendibal takes on Sura Novi as a companion. In "Stor Gendibal, Sura Novi" we find: "I've been listening to U, R.A. Leedi. I believe U are going to be our leader and savior." Sura Novi's full name is "Suranoviremblastiran." In this we can find: "I, Asimov, listen to U, R.A.L. U believe U mite soon be our savior." In "Blissenobiarella" (34) we find: "I be a lissener as R.A.L. labors." Another member of the Second Foundation's governing table is "Delora Delarmi," whose name may contain this message: "Am I a moral leader or merely a riddler?"

Getting back to C.T. Tincher, in "Chetter Hummin" (66) we can find this message: "Eech time U meet C.T. Tincher I C him enter U, Ruth." On this same theme, in "Kelaritan, Cubello, Sheerin" (120) we find Asimov observing as Clay instructs me about sex: "I listen, I look as C.T. Tincher teaches U to be better at secs, Ruth Anne L." This message is confirmed by a similar message that can be found when a fourth character joins this group. In "Kelaritan, Cubello, Varitta, Sheerin" we can clearly see "lovar." The message appears to be: "I listen, I look as C.T. Tincher teaches U to be a better lover, Ruth Anne L." From the same novel, in "Sagikan, Beklimot, Thombo" we find: "I am both looking an' listening. It is a gamble." Asimov spends time observing me and preparing riddles for me to find, to help prove that he was observing me because of my cataclysmic writing, and all this could be a waste of time. It is a long shot, to gamble that one person will survive and find his riddles, decipher them accurately and publish them in an attempt to end a cover-up. If that one person does not come through, all that effort has been wasted.

As Clay's interest in me waned, his friend Richard Johnson came into my life. Both had a military background. Rich could not qualify as a pilot on account of his eyesight, but he took me on a hair-raisingly fast test ride in the Ford EXP which I bought around 1989. In that year I became friends with John W. Carr, to whom I am now married. In "The Smile of the Chipper" (35) Asimov gives three character names: "Johnson, C, F." This is cryptic, but the message could be: "Johnson shos off. John C. no shos off." A couple of other riddles that appear to mention John have been given in the preceding chapter. My friendship with John did not take a romantic turn immediately, partly because he was such a decent guy and partly because I had promised my new landlord Sam Wilson that I would take up with him if I broke up with Rich Johnson. Before that break-up occurred, Sam seized a chance to pin me down. I had caught a rat in a trap and took it up to Sam's house (from my mobile home), to see about disposing of it. Sam invited me to look inside his hunting lodge, which was a mobile home behind the house. On the wall were racy cartoons. All of the sudden he was on me, very strong, and what the heck. I resisted, but it was no use.

In "Hallucination" (42) Asimov presents a character named Sam Chase. He has conversations with four other characters. In "Sam Chase, Donald Gentry" we find: "Yoo enter landlord Sam's lodge, Rooth Anne Leedy. He chases yoo, gets on yoo. Yoo try to dodge him. Yoo can't." In "Sam Chase, the Corridor-Master" we see: "Sam chases Rooth, rides her. This is a crime." After my break-up with Rich I did agree to an affair with Sam. In "Sam Chase, Robert Gillettee" we can find: "Sam chases Rooth L., grabs her. Later she lets him get her, agrees to be his girl." I broke off with Sam, telling him his lovemaking was too rough, not comfortable. He tried unsuccessfully to interest me in another lover. Then one day as I spoke on the phone, he came into my mobile home and waved his male member in my face a while. When I hung up, he attempted to corner me and almost succeeded. I went outside, and he came out and stood there asking me whether I wanted him to stay or go. As tactfully as I could I asked for him to go. Thereafter he married another woman.

In "Sam Chase, the Commander" we can find: "Sam enters Rooth Anne's home, shos a hard-on, chases her, demands secs. She heads to the door. He can't catch her. He stands near her, ascs shood he sta or not. She ses no. He comes to her no more."

After Sam Wilson, John Carr was my only boyfriend. We bought a house together in 1992 and were married in 1995.

About my two amorous landlords, I want to say I did not feel I was low or sleazy when involved with them, or that they were low or sleazy for pursuing me. One doesn't know whether a relationship will last or not. My main feeling was that I was grateful to each of them for offering me a lifeline in the form of affordable rent in a country setting, which was what I wanted. These rentals enabled me to avoid situations that could have been disastrous, on account of mothballs, gas fumes, car exhaust and other problems. I was kind of exhausted, and I was lucky to find places to live where I could begin to thrive again. Yes, I was grateful to my landlords.

If I appeared sleazy to others, look at it from Isaac Asimov's point of view. He needed someone to help spread his message regarding hollow Earth and cataclysm. I was the only person offering a lifeline to him in his quest. He was not too proud to seek my help, even though I had hit the skids, not too proud to continue to fashion riddles designed to help me end the cover-up. He accepted my help gratefully. There was no one else.

9.
"PLANET EARTH BE HOLLOW. NEW ROTATION PATTERN WILL RECARVE EACH POLAR HOLE." - ISAAC ASIMOV

The strongest hollow Earth statement we can find in Isaac Asimov's work is the double message we have seen in "Perfectly Formal" (45). The message "Planet Earth be hollow" can be found in "Winthrop Carver Cabwell" and also in his fiance's name "Hortense Hepzibah Lowot," with "Planet Earth is hollow" also being present in her name. None of Asimov's other character names contain this exact message, spelled correctly, and yet it comes up in both these names.

An axis shift message can also be found in "Winthrop Carver Cabwell." The middle name strongly suggests "recarve." The recarving of polar holes due to centrifugal force as the axis migrates is the most alarming feature of an axis shift, although the eruption of large and small volcanoes will also be a major threat to many parts of the world. The sinking of Japan and other areas near the new equatorial bulge will also be a major concern. Let us look at what Asimov appears to be saying in "Winthrop Carver Cabwell:"

"Planet Earth be hollow. New rotation pattern will recarve each polar hole." In a name only 21 letters long, it is remarkable that Asimov has given us a long and detailed message with every word spelled correctly. This message is not only very long and very clear, but it also is the most vital message he is trying to get across regarding a coming cataclysm. This tells us, better than anything else he could say, exactly why an axis shift is going to be so horrible.

A similar message can be found in "Hortense Hepzibah Lowot," as follows: "Planet Earth is hollow. New rotation pattern will reshape the polar holes. When people realize this it will be possible to prepare." Many preparations will require government action and cooperation between governments, and so these preparations will not be possible until the truth is generally known around the world.

"The Earth shall be shaken as poles are reversed and polar holes rekarved." This is from "Bel Arvadan, Pola Shekt" (65). In "Alexander Hoskins" (36) we find: "In six dekades all land near axis holes shall shake." Similarly, in "Hari Seldon, Agis XIV" (30) we can find: "All hoo live near axis are in danger as a reversal nears." In "Louis Peyton" (105) we find: "You'll see nu poles in nu positions soon." This character changes his position dramatically, from Earth to the moon and back.

"In mani reversals, axis remains (the) same." This message from "Maximian, Meliversa" (56) tells us that if the north magnetic pole migrates to the same location previously occupied by the south magnetic pole, and vice versa, the solar wind will not define a new axis as a result of a geomagnetic field reversal. The expectation that the coming reversal will bring a new axis may result from the fact that all land masses have been drawn northward in recent geologic time (141), to the point where the north magnetic pole must head for an area surrounded by more water.

"Geomagnetik field rearrangement to kome after my time." This is from "Gregory Arnfeld, Tertia, Mike" (76).

The word "shift" can be spelled from "Edith Fellowes" (122). Her journey back in time to join Neanderthal society can be seen as a symbol of our own civilization's coming return to a primitive state if we do not confront the threat of a coming axis shift. Through her name Asimov may be saying: "The holes to the hollow of hell will shift." One could add: "Deths will follow." Likewise, in "Isidore Wellby, Shapur" (94) we find: "I say our world is hollow, will be reshaped. A polar hole will be where upper Europe is." Asimov may have been urging me in one of his final stories to renew my warnings about axis shifting. After doing five books between 1981 and 1988 I was doing shorter, less stressful projects. In "Euphrosyne Stump" (50) he may have been saying: "Soon you must prophesy some more to Europe, Ruth." When we include Euphrosyne's husband "Alexius Mellon" in the riddle, the message may be: "Ruth Anne L., you must prophesy more to Europeans on a shipht in the holes at the planetary axis." Several more riddles about how an axis shift will affect Europe have been seen earlier. The "fantastic voyage" within the human body, as described in Asimov's 1987 novel (29), takes place entirely within the head of a dying patient, "Pyotor Leonovich Shapirov." His head seems as large as Earth to the travelers in their miniaturized state. This comparison between Shapirov and the Earth helps confirm the message we can find in his name: "A pole reversal shall reshape the Earth catastrophically in seventy to ninety years." Similarly, in "Valeri 'Ashby' Paleron" (29) we can find: "As yoo prophesy, R.A.L., a pole reversal is probably very near." In "Janov Pelorat, Bliss" (34) the message may be: "A pole reversal is liable to jar planet soon." One could add: "Japan is to be lost." In "Jirad Tisalver, Casilia" (66) we see: "A reversal is a veri larje, veri scari disaster." In "Stettin Palver, Joramis" (30) we can find: "Planet is to see a major pole reversal sometime soon." In 1951 the name "Poly Verisof" (31) seemed to say: "Reversl of poles is severel yeers off." Verisof assumes a prophetic role in the novel. In "Jonas Foster, Ralph Nimmo" (90) we see: "A major polar shift is imminent on planet Earth."

ERUPTIONS AND FLOODING—THE EQUATORIAL SHIFT

Since the Earth is oblate, the shifting of its bulging equator is bound to cause disruptions—changes in sea level and seismic activity. We recall that Edgar Cayce predicted eruptions of volcanoes in the "Torrid area" as well as polar upheavals to occur in conjunction with a shift in the planetary axis.

Isaac Asimov appears to have expressed both an axis shift and volcanic activity in tropical areas in a single riddle. The character name he uses belongs to the president of Earth in *Nemesis*. This individual represents Earth politically, which makes it easy for us to accept that he also is a symbol of Earth itself. He has a temper that "explodes like Krakatoa," which is the name of one of the best known violently active volcanoes. In "Igor Koropatsky" we can find: "Aksis is to tip. Sky at tropiks is to gro gray as Krakatoas go pop." Another president of Earth from *Nemesis* also has "Krakatoa" in his name. In "Kattimoro Tanayama" Asimov may be saying: "Rotation to tak a noo tern. Many Krakatoa at a tim ar to mak kommotion."

The supervolcano Tambora is strongly suggested in the name "Tambor Brossen," also from *Nemesis*.

Brossen has also been director of a world, but it is a small manmade settlement that travels through space in search of a new place to settle. In this name we can find: "Tamboras, Tobas are to be reborn soon." His successor is named "Janus Pitt." Adding the name of the settlement he directs, which is "Rotor," this message can be found: "Japan is soon to turn into a ruin." The ruins of Japan will be underwater, swamped by the new equatorial bulge, if the planetary axis migrates to northern Europe. Such a change would also bring high water to Hawaii and the western United States.

We have already seen riddles from Asimov predicting trouble for these areas. Cayce predicted that parts of the western United States would break apart.

DEFINING THE EARTH'S HOLLOW STRUCTURE

"In the Earth's interior is a vast cavern. At its core there shines a son." This is from "Henrietta Corvass" (62). Asimov even tells us, in "Littoral Thoobing, Sayshell" (34), why the hollow shell does not collapse due to gravity: "I say the Earth has a hollo shell that orbits a little star at its senter." In "Wilson Roth" (114) he gives a simple, direct statement: "This Irth is hollow. Within it is a son." In "Roj Nemennuh Sarton" (92) we find: "Earth has a small sun at senter that's not as huje as Earth's Moon." In "Albert Jonas Morrison" (29) we can find: "Inner son is not as larje or as brite as Moon." In "the Estwald theory" (108) Asimov encourages us, through the word "theory," to look for "hollow Earth." The message may be: "The hollow Earth theory does hold water." The traditional view of a hollow Earth, as portrayed in myths, appears to be reflected in the message we find in "Harley Smythe-Robertson" (121): "There are many early myths that say the Earth be hollo." In "Charlie Janowitz" (44) we find: "I, Izaac A., tell ewe that the Earth haz a larje inner hollow." Likewise, in "Zelda Charlowicz" (87) we find: "Izaac A. haz decided world iz hollow." There are alternate solutions to be seen later for the Charlowicz and Janowitz puzzles. In "Pelleas Anthor" (80) we see: "There are three hollo planets, Earth an' too others." Asimov tells us in "Corwin Attlebish" (113) how he feels about the secrecy surrounding the hollow Earth issue: "As a science writer I want to be able to tell others that the Earth is hollow." One could add: "With brain control I cannot." The phrase "win battle" is also possible. The Holy Grail may be included in the message we find in "Anthony Gerrigel" (92): "In Holy Grail lore the Earth hath a large inner hollo." In "Elizabeth Thornbowe" (92) Asimov makes it clear that he would reveal the Earth's shape if he could:

"I be not able to tell ewe that the Earth iz hollow." One could add: "Brain iz bent." We can see "bent" clearly in "Andrew Harlan, Noys Lambent" (107), where the message appears to be: "Eye am aware that the Earth be hollow, and yet Eye am not able to say that to anyone. My brane has been bent." In "Matthew Hortenweiler" (90) the message appears to be: "I want to tell ewe that Earth an' Mars are hollow." In "Emmer Thalus" (66) the message may be: "The Earth has a small star. U must trust me to tell U the truth." In "H. Seton Davenport" (105), who has a star-shaped scar, we can find this message: "To save the Earth, one needs to prove to others that Earth has a star at senter."

"...MY ANAGRAMS MAY BE NECESSARY"

Evaluating anagram messages is no picnic. I have done many puzzles over and over, then gone back and changed my interpretation. Almost any riddle message can be shrugged off as possibly being wrong. This is why I have presented so many riddles on each major point. Now let us look at some riddles about riddles themselves.

"Secrecy means my anagrams may be necessary." We find this message in "Cambyses Green" (60). This character calls Earth a "hellhole," which suggests that the secrecy expressed in his name may relate to the existence of a real hell, plus a hole leading to that hell.

"Meet the real Riddler. I am he." This is from "her middle-mate" (51), a term from Asimov's stories featuring the demon Azazel. From another of these stories there is "the dread Lamell Bird" (7), where we find: "I am labelled 'the Riddler.'" In "Oliver Heaviside" (57) there is this: "I have loved role ov savior, deliverer." In "Elderberry Muggs, Hank" (9) we see: "Yu kan learn my ryddle game, R.A. Leedy. Yu shall be my messenger." In "Kelden Amadiro" (114) we see: "I'll enkode married name o' Kallinan and maiden name o' Leedi in mani a riddle, R.A. Leedi." In "Rikaine Delmarre, Gladia" (113) we can find: "I am a real riddler, am making mani a riddle." In "Daneel Giskard Baley" (79) we find: "R.A. Leedy is a real brain, kan learn riddle reading." In "Gebore Astinwald" (79) we see: "I want riddles to be read to world. I don't want ewe to be in danger, R.A. Leedi." In "Richard Youngerlea" (73) we see: "R.A. Leedy, you are young and hardy. You can learn riddle reading." In "Hari Seldon, Yugo Amaryl" (30) the message may be: "R.A. Leedy, I am sure you shall do my riddles someday and share my messages on doomsday." In "Chester Dunhill" (68) the message may be: "Ruth Leedi, U did see hints. U did see clues. U didn't see these riddles."

In "Dugal Tennar, Render Linn" (30), the names of two autocrats, there may be a cautionary message: "I didn't intend to endanger U, R.A. Leedi, thru ani riddle that I did." In "Ludovic Broadbottom" (68) Asimov may be saying: "U ar abl to do a riddl, to rit a booc, to B mi voic, R.A.L." In "Hari. Seldon, Raych" (30) we see: "R.A. Leedy, you share Isaac's hints. Share his riddles also." In "my poor middle mother" (12), Asimov's message may be: "I'm the reel Riddler. I hope I mite help yoo promote the hollo Erth theory, Rooth Leedy."

THE STRONGEST RIDDLES WILL REMAIN VALID

As you are confronted with a blizzard of riddles, I ask you always to bear in mind the most powerful ones, which were given in Chapter 1, as well as the messages adjacent to punctuation, which prove that secret messages were given. Considering individual names from Asimov and Sagan, forming riddles on hollow Earth or Mars and/or an axis shift, I would list the tollowing as the most powerful:

1. "Alfred Pennyworth" (53) contains the letters for "hollow planet Earth" plus "world" and "theory." One of his fellow characters in the Batman franchise is called "the Riddler," which tells us that riddles are likely to be found among these character names.
2. "Stefan Alexeivich Baruda" (118) contains the letters for "Earth axis shift" and "field reversal." The word "decades" can be made, allowing Sagan to indicate how soon a field reversal might lead to an axis shift. This character gives a doomsday warning.
3. "Hober Mallows" (34) easily converts to "Mars be hollow."
4. "Hor Devallow" (65) easily converts to "We have a hollow world."
5. "Winthrop Carver Cabwell" (45) contains the letters for "Planet Earth be hollow." This is remarkable even without his fiance having the same words in her name.
6. "Hortense Hepzibah Lowot" (65) contains the letters for "Planet Earth be hollow" as well as "Planet Earth <u>is</u> hollow." This is an outstanding pair of messages, even without her fiancé having one of these messages in his name.

These riddles are strong evidence for hollow Earth and a coming polar shift, regardless of any doubts you may have about the dozens of other messages you see here. Also, even if the complete message I suggest for one of these puzzles is incorrect, the presence of powerful words in these names is still extremely significant.

10.
POLAR HOLES ARE DETERIORATING ON MARS AND VENUS; "MARS IS ABOUT TO GO LIMP, CRUMPLE." – ISAAC ASIMOV

When Isaac Asimov took the risky step of placing three names from *Nemesis* next to punctuation in his story "Frustration," he was driven by an existential threat to Earth. Both his novel *Nemesis* and his story "Frustration" concerned the possible destruction of Earth, either by the approach of a secret star or through nuclear annihilation.

Why was it extremely risky for him to encode the names Merry, Tessa and Ranay into the letters before and after commas and apostrophes in his story "Frustration"? If it hadn't been risky, he wouldn't have used a code at all. He would have made a clear statement in an article or in a news conference. It is easy to tell that he felt the need to be subtle. Most people would never dream of writing down all the letters before and after commas and apostrophes in a story. And yet once the message is found it cannot be denied, because all the letters are in the proper order. Anagram messages, which require rearranging the letters and perhaps reusing some letters, can always be shrugged off as the figment of someone's imagination. But there is no denying the Merry-Tessa-Ranay message. And so if Asimov was defying a powerful cover-up by creating that message he was risking retaliation against himself, his family, and his editors and publishers. And anyone who finds the message and shares it is also taking a risk.

That risk is reduced somewhat by the vagueness of the message. One wonders whether these names formed a riddle, or whether the behavior of these characters offers some clue as to what Asimov was getting at. But his purpose is pretty simple, when you think about it. He wanted people to look hard at the novel *Nemesis* for a portrayal of a real crisis on Earth.

Could a secret star really be heading for Earth, as portrayed in the novel? A close reading makes it clear that a star that has been kept secret from us really could threaten this planet. The "star" of the novel is the heroine Marlene Insigna Fisher. Her name begins with "Mar," which suggests the term "Mar-star" or even "Mars star." Does Marlene, the Mar-star, head for Earth? Actually she does. This behavior doesn't make a lot of sense, but of course symbolically it is perfect for sending a meessage about the Mars star heading for Earth.

Marlene is a member of a settlement that has left the solar system to establish itself around another star. Marlene's mother Eugenia is an astronomer who discovers that a distant star is heading for the solar system. Her boss persuades her that this threat should be kept secret from Earth. The settlement then proceeds to establish itself in the most dangerous possible place, in the area surrounding the Nemesis star that is heading for Earth. There is no way that this can be considered a realistic choice. Rational people would have chosen to settle near a star that was on a safe course, and there would have been thousands to choose from. Somehow Asimov contrived to have these people choose the one destination that would allow him to portray the "Mar-star" heading for Earth.

On arriving near Nemesis, Marlene proceeds to bond with the moon-world Erythro, similar to Earth, which others are afraid to explore. She decides to remain permanently on that moon-world near Nemesis. The entire star system is heading for Earth, where it will disrupt Earth's orbit and render it uninhabitable, but the worlds near Nemesis are "more tightly held" by gravity. Being more tightly held, they will not be disturbed by the encounter which will be so damaging to Earth. This sounds improbable. If Earth's orbit stands to be disturbed, then the orbit of Erythro should logically be disturbed by the coming encounter as well. In the end it doesn't matter because a plan is devised to divert the Nemesis star away from the Earthly solar system. That plan is also improbable, but the novel makes its point. The Mar-star Marlene makes a decision to remain near the secret (or formerly secret) star Nemesis as it heads for Earth. So the Mar-star is heading for Earth.

The similarity of Erythro to Earth must be intended to help portray the attraction between the "Mar-star" and Earth. No one else on the settlement dares to explore the moon-world's surface, but she goes out there repeatedly, being irresistibly drawn to do so. If the "Mar-star" has an unusually strong tendency to bond with an Earth-like world, this suggests that the Mars star, after that planet's collapse, will be drawn through gravity to bond with planet Earth.

In a rather obvious way Asimov has used the star of the novel to portray a coming threat to Earth from a star. She is not just any star. Her name tells us she represents the star in Mars. What Asimov has done is similar to what writers have done for centuries to make subtle statements on controversial topics. If this symbolism is rejected by readers, there is no other obvious purpose to be served by Asimov's secret message consisting of the names Merry, Tessa and Ranay—three characters from *Nemesis*.

The diversion of the Nemesis star to a new path, as improbable as it may sound, tells us that the inner star of Mars, or in effect the entire planet, can be diverted onto a new path by us. If we act soon enough, that new path will protect Earth from impacts in the future, after hollow Mars falls apart.

If "Marlene Insigne Fisher, Erythro" is treated as a riddle, the message may be: "In fifty to eighty years the star in Mars shall threaten Earth." Before Mars falls apart it will sag limply for a period of time.

"Mars is not yet (totally) limp." This message from "Limmar Ponyets" (31) tells us that as early as 1951 Asimov was aware that Mars was destined to fall apart due to lack of solar wind support for its rotation. The orbiting shell relies on centrifugal force from rapid rotation to hold it in its position in orbit. Without adequate rotation speed the outer shell of Mars will collapse due to gravity. It has been thousands of years since Mars had a north-south magnetic field which served to funnel the solar wind's force to the north and south magnetic poles. That concentrated force, striking at the ideal distance from the axis, was sustaining the rotation of the entire planet. Now that the north and south magnetic poles have died away, the planet has been coasting in its rotation, with no boost whatsoever, for thousands of years.

"Mars is about to go limp, to vibrat, crumpl, collaps." This message can be found in "Alistair Tobago Crump VI" (11). In "Mary Ann Lipsky" (62) we find: "I say, as plainly as I kan, Mars is limp." In the same novel, in "Martin Walters," we can find: "In time Mars will release its internal star." The name Marsogliani, which we saw earlier, also comes from this novel. With names beginning "Mar" and "Mar" and "Mars," this novel, entitled *Murder at the ABA*, seems to have a special focus on riddles about Mars. "Michael Strong," which is another name from this novel, can be expected to carry an additional message about Mars, even though the letters for Mars are not grouped together. The message here may be: "Mars rotation is stalling. Its central star is a threat to Earth." With all these alarming messages about Mars, this novel can be considered to be sending an SOS regarding a threat to Earth from outer space. This type of situation is also portrayed in "Old-fashioned" (95), where SOS signs are actually sent to Earth. Since the SOS sign is an ABA form, as we saw earlier, the fact that "ABA" is found in the title of this novel strengthens the impression that the novel is sending an SOS to Earth. The name "Nellie Griswold," also from *Murder at the ABA*, appears to convey this message: "End o' world riddles need reeding. No one is willing." This also adds to the impression that the novel is sending an SOS. In "Thomas Valier," also from the novel, we can find: "Tri to move the Mars star or it shall hit the Earth." This message resembles what we can see in "Leander Thomson" (17): "Do not let the Mars star land on Earth." A large orange ball (resembling a sun) lands badly in this story.

"A star at Mars senter wants to wander soon." This is from "Andrew Mortenson" (5). In "Mordecai Sims" (19) we find: "Mars is doomed. I am scared." In "Mandress" (34) there is this: "Mars end nears."

HOW "MARSHALL" TELLS US MARS SHALL END

The most chilling predictions about Mars in the riddles of Isaac Asimov and Carl Sagan come from riddles containing the name Marshall or the word marshal. These messages are chilling because we can see the words "Mars shall" right before our very eyes.

When Carl Sagan mentioned "the Marshal Nedelin" (148) he was making it easy for us to see the message "Mars shall end."

Transpose the first two letters of "Nedelin" and you have "end" at the beginning of the name. The entire message, as we have seen, appears to be: "Mars shall end and its inner star shall threaten the Earth." This is powerful, but I do not see any related messages nearby which would help to confirm this message.

When Isaac Asimov presented "Marshall Sebatinsky" in a riddle, a series of related riddles were also present (88). The name appears to carry this message: "Mars shall break. Then its inner star shall be seen in the nite sky as it threatens the Earth." When we add the name of Sebatinsky's wife "Sophie" to the riddle, we can find: "Hollo Mars shall kollapse. Then its inner star shall be seen in the nite sky as it threatens to hit planet Earth." This dangerous situation is described again in the name "Mikhail Andreyevich Zebatinsky," where we can find: "It is necessary that every citizen realize that Mars shall break, and then the star inside Mars can easily strike the Earth." A commentary on this danger can be found in two character names that go together, "Henry Brand, Paul Kristow." In these names we see: "You don't want hollow planet Earth to be struk by a star or an asteroid." A follow-up to this can be found in another pairing of characters: "Henry Brand, Albert Quincy." In these names we find: "Yu'll see the threat and yu'll be required to rac yur brain(s) in a hurry to handle it." Also from this story, in "Haround, Mestack" the message may be: "U must understand that others can use the Mars star to attack the Earth." These two characters are not ordinary humans but rather a pair of aliens observing humans from some other plane of existence.

ARE POLAR HOLES DISINTEGRATING ON MARS AND VENUS?

Increasing limpness may not be the only sign of instability on Mars. Some disintegration seems to be described by Asimov in "Simon Alexander, Johnson" (27), as follows: "Mars is doomed. Land near axis no is normal. Small losses are seen. Soon one shall see larjer losses." A similar abnormality near the axis of Venus is described by Carl Sagan in two riddles. In "Vasily Gregorovich Lunacharsky" (148) we find: "Holes on Venus are changing, groing larger each year, very likely cracking." In "Anatoly Vasilyevich Lunacharsky" we see: "I check holes at the aksis on Venus annually. Every year the holes are a little less circular, less even. To check on this is very risky." We have seen a threat from Venus described by Asimov in "Ben Estes, Harvey Funarelli" (95), as follows: "I fear Venus's shell is unstable, shall release its little sun in fifty to seventy years." The threat from Mars is summed up this way by Asimov in "Laskin Joranum" (30): "Mars sun is a jinks on us all in a loos kannon." If the equator of Mars were constricted, the planet would be like a cannon with a cannonball halfway down its barrel. Another warning about Mars can be found in "Benjamin Allan Denison, Andrew" (108), where we see: "Be warned. World is in danjer. Small sun inside Mars will soon wander." In "Lamar Swift" (75) the message appears to be: "Mars star will fall swiftli." Sagan provides this possible message in "Sol R. Hadden, Yamagishi" (148): "Inside Mars a hidden son, or sol, is shining. Share messages on doomsday hidden in my riddles, R.A. Leedy."
In *Nemesis* Asimov presents a doomsday scenario in "Siever Genarr, Erythro Dome." as follows: "Mars is doomed, and its inner star is on the verge ov destroying the Earth in seventy years." Likewise, in "Merrill 'Handles' Foster" (4) we find: "If the star inside Mars falls on Earth, the Earth dies."

A strangely complex character name in *The Norby Chronicles* (119) would almost have to be a puzzle because it is unpronounceable. In "Ziphyzggtmtizm, Zargl" we can see the letters for "Marz" as well as larg, graz, iz, iph and itz grouped together. The message appears to be: "Marz iz limp. Itz gapz (polar gaps) ar larg. Iph Marz phallz apart itz partz may graz Irth, may hit Irth." The names in the puzzle belong to dinosaurs, who are reached through time travel. Since dinosaurs were wiped out by asteroid impact, their presence here helps to confirm that this puzzle is about a potential asteroid impact. In "Hober Mallow, Asper Argo" (31) we see: "Mars be hollow. Polar holes are large, grow larger." Likewise in Genovus Pandaral" (79) the message could be: "Polar gaps on Venus are enlarged."

CAN THE BREAK-UP OF MARS BE PREVENTED?

Two possible solutions for the Mars threat seem to be suggested by Asimov. We have seen how the stellar threat was resolved in *Nemesis*, by altering the course of the star that was heading for the solar system. This suggests that we should push on the Mars star with bombs in hopes of sending the entire planet onto a course that will not threaten us. The other possible solution Asimov may be suggesting is in "Brock, Ames" (85), where we find: "Make Mars core come back." One could add: "or Mars breaks, makes a mess o' rocks." This suggests that the planet's core, which used to generate its north-south magnetic field, could somehow be re-energized. Perhaps our planet's ruler could be persuaded to restore it, if he can.

The phenomenon of sunlight on the shaded winter pole of Mars has been discussed here and shown in photographs. Asimov appears to have used the character "Maurice Winters" (71) to provide his own discussion on this topic. The message here appears to be: "An inner sun causes martian 'ice' in winter." This tells us that the phenomenon of a brightly shining so-called ice cap at the winter pole of Mars actually comes from the light of the planet's inner sun.

"Mars sun hydes no more. Rays are seen." This is from "Seymour Norman Hyde" in Asimov's story "Gift" (117), which was published shortly after I sent him my 1981 book. From the same story, "Morris Quintus Yeats" may contain this message: "Sientists must stay quiet as to a star in Mars." The word "visible" is strongly suggested in "Elvis Blei, Steven Lamorak" (89), where the message may be: "Mars inner star's lite is visible to me, Isaak Asimov. I am easili able to see it." Likewise, in "Beenay, Raissta" (120) we find: "An inner star is seen by sientists." Beenay is an astronomer. The Mars threat is seen again in "Harrim, Trinit, Sheerin" (120), where the message may be: "In time Mars shatters. Then the star in Mars mite hit the Earth." Likewise, in "Ishtar Mistik" (15) we find: "Mars star mit mak a mistak, hit Irth." In "Thaddeus Araman" (90) Asimov could be saying: "The Mars star means sudden death tu us, tu the Earth." In "Aurinel Pampas, Dolorette" (63) the message may be: "People must understand, a star in Mars puts our planet in peril." A plan to wipe out humanity is devised by "Martinus Augustus Dander" in "The Mad Scientist" (57). In this name Asimov may be telling us: "U must understand, a star inside Mars is a danger tu U."

MORE ON ASIMOV'S MESSAGES BESIDE PUNCTUATION

So far we have looked at the "Merry-Tessa-Ranay" message which Asimov hid next to commas and apostrophes in his story "Frustration." The three names were evidently an indication that the novel *Nemesis*, from which they came, contains an important warning for us regarding a stellar threat to Earth. We have not looked in detail at the message "lestwe al die, die, die," which was hidden next to periods in Part IV, Section 26 of *Forward the Foundation*. There was a third message which we have not looked at yet at all. I am sure there are other messages that I have not discovered.

The phrase "lest we all die, die, die" sounds like it is meant to underscore a warning given earlier. For example, "Merry, Tessa, Ranay" could be a riddle containing a warning or a piece of advice Asimov wants us to follow, for our own protection. In studying these names I see two possible messages that would make sense if placed before the phrase "lest we all die, die, die." The combined statement could be: "Steer Mars star. Try many years, lest we all die, die, die." Also, the completed statement could be: "Ye may enter Mars, steer Mars star. Eye say try, lest we all die, die, die." This second choice seems to have some rhyme to it, which would be appropriate for an author who has done a lot of rhyming poetry. I think the message should be about the Mars threat because the three names come from a novel about a threat from a secret star, with hints pointing to Mars as the source of that star.

The third message beside punctuation comes from the two sections of *Forward the Foundation* where the tiger and camel (greti and lamec) anagrams can be found. The punctuation involved was commas and apostrophes, as I recall. The words that emerged in one section included "war" and "head" and "Urth." The presence of "Urth" is extremely significant because Asimov wrote four mysteries featuring a Dr. Wendell Urth (98, 103, 105, 106). The most significant word that emerged in the other section seemed to be "Murs." Since we just saw "Earth" respelled as "Urth," it should not be surprising to find "Mars" respelled as "Murs." Seeing this emphasizes for us how dangerous it was for Asimov to make a clear and explicit statement about the Mars threat. The combined message seems to be that Murs (Mars) is a warhead aimed at Urth (Earth). In other words, Mars has been weaponized against us by enemies that we are scarcely aware of.

The message in "Wendell Urth" may be: "We need new leeder tu leed the Erth." If our planet is on a collision course with pieces of another planet, and if it is all a big secret, I'd say yes, new leadership is needed. Taking another look at "Merry, Tessa, Ranay as a riddle, two additional words may belong with the three names. The word "Celts" can be found in the list, and if letters adjacent to single quotations marks are included int he list, the word "die" can be found. In "Merry, Tessa, Ranay, Celts die" the message may be: "A cataclysm is near,. Celts and Danes and all near Denmare may die."

11.
"ASK U.S. PRESIDENT TO CONTACT OUR PLANET'S RULER"; "AZAZEL IS IMMORTAL...RULES UNSEEN"

The ancient Hebrews believed in the existence of a powerful individual named Azazel. This is revealed in Isaac Asimov's introduction to his 1988 story collection entitled *Azazel*. Each story portrays Asimov himself as listening skeptically to a friend's description of an incident where a two-centimeter demon grants a wish without solving any problem. The stories are presented as fantasies, but in view of the hints and riddles in Asimov's work, I have suspected that Asimov was trying to convey that there was a real Azazel who was the mastermind of a cataclysmic cover-up.

One tipoff that Azazel might be real is the character name "Lucius Lamar Hazeltine" (55) from one of these stories. We can see the letters "azel" showing in this name. The message may be: "Azazel rules the Earth, silences all scientists, harms us." Other messages are possible here, but this is an important statement telling us who Azazel is and how he operates. Hazeltine's fiancé is "Agatha Dorothy Lissauer," in whose name we have seen this message: "You say the Earth is hollo, Ruth A. Leedy. You are right." Another possible message is: "Asasel says he is going to use a star to destroy the Earth." This solution may be the better of the two because "hollow" is misspelled in the first solution we have looked at. Also, the second solution seems to be more clearly related to the message we see in the name of Lissauer's fiance Hazeltine. The idea of a star possibly destroying the Earth is the theme of Asimov's novel *Nemesis*. He emphasized the importance of this theme by encoding three character names from the novel into the letters before and after commas and apostrophes in his story "Frustration" (40).

Hazeltine and Lissauer come from a story whose title contains "Earth." In "The Critic on the Hearth" we find: "Rooth Anne can teach the correct theori on the Earth. I cannot."

There are two z's in "Mitza Lizalor" (32), and so I have long considered this name to carry a message about Azazel. I believe the name could be telling us: "Azazil iz immortal." Another name with two z's in Asimov's work is "Zelda Charlowicz" (87). The message here may be: "Azazel haz decided which hollow worldz will collide." In "Brutus Franklin Hunzinger" (70) the message may be: "You are fearless, Ruth Anne L. Surely you realize Earth's king Azazel is a brute, kan break your brain." In "Fifi Laverne Moskowitz, Sophocles" (10) we can find this: "Few people know of the hollow Earth's ruler Azazel. I, Isaac Asimov, know how to call him." Likewise, in "Merrilee Augina Blankowitz" (63) we find: "I know ruler Azazel'z language. I am able to talk in it."

In "Lanel Russ" (66) we find: "Asasel rules unseen." If his partner's name "Gebore Astinwald" is added to Russ's name, the message appears to be: "Asasel rules world unseen bi bending and tearing brains." In "Las Zenow" (30) we can find: "Azazel nose ewe." Adding Zenow's home world "Wencory" to the riddle, the message may be: "Azazel nose ewe were once crazy, R.A.L." My breakdown at age 16 involved romantic and religious fantasies. In "Golan Trevize" (34) we can find: "I long to reveal all I no ov Azazel."

When the name of Trevize's girlfriend "Flavella" is added to his name, we find: "I long to reveal all I no ov Azazel'z love life." When his lover "Hiroko" (32) is added to his name we find: "I long to reveal that Azazel iz king ov the Earth."

When Trevize's name is combined with that of his lover "Mitza Lizalor" (32), we see: "I long to reveal Azazel iz immortal."

Carl Sagan appears to touch on cannibalism in a message derived from "Billy Jo Rankin" (148): "Royal jinn B a kannibal." One has to suspect cannibalism when news of a cataclysm is suppressed. I speculated on that in my first book. The noted UFO researcher Charles Fort suggested that Earth may be a gigantic farm, where people are urged to be fruitful and multiply and are also groomed to anticipate destruction.

Mass murder is also suggested in Asimov's riddle "Alban Wellis" (66), where we see: "All will be slain." This riddle appears shortly after "Osterfith," where the message appears to be: "The Erth is to shift."

AZAZEL IS DESCRIBED AS APE-LIKE

Different ideas about Azazel have seemed right to me from time to time, based on my early efforts with riddles. Azazel was a cyclops. He was a centaur. He was a Sasquatch.

"The big sasquam" is a made-up term from one of Asimov's stories about the demon Azazel. That phrase is spoken by Azazel, and is used as an epithet. "Sasqua" sounds like the beginning of "Sasquatch," and no other possibility comes to mind. The word as a whole sounds like a combination of "am" and the beginning of "Sasquatch." It looks as though Azazel is portrayed as telling us: "I am a Sasquatch." The use of the word "big" here reminds us of Bigfoot, which is another name for Sasquatch. In "the big sasquam" (60) we can find: "I am a big Sasquatsh. Mi teeth be big." Other possibilities are: "I am the biggest Sasquatsh," or "U must be quiet as U meet the big Sasquatsh." Similarly, in "quackenbush" (79) we can see: "Sasquach can cause quakes. U can be shaken." In "Merton Mansky" from "The Bicentenniel Man" (99) we can find: "A monkey-man's my master."

In "The Monkey's Finger" (104) Asimov portrays an unusually intelligent monkey being able to type out sentences that are understandable, though the spelling is imperfect. The character names in this story are "Marmaduke Tallinn" and "Lemuel Hoskins." Both names contain the word "monkey," but with imperfect spelling. In "Marmaduke Tallinn" we find: "A tall dark munkie-man ruleth me." In "Lemuel Hoskins" the message may be: "Monkie-men in hell's hollo no look like us."

The very fact that Asimov portrayed the demon Azazel as having a tail lends some support to the idea that the real Azazel is one of the great apes. The fictional Azazel is two centimeters tall, which is obviously a fantasy. But this may be meant to tell us that size is an issue with the real Azazel, that in fact he is quite tall.

WHY WOULD AZAZEL TOLERATE MY HOLLOW PLANET PROJECT?

With my early books it seemed inevitable that there would be retaliation against me. If thousands of scientists were prevented from revealing that the Earth is hollow, why should I be allowed to speak freely on the subject and put forward photographic evidence? I could imagine the mighty Azazel belittling my work, saying I was not a real threat to the cover-up, but at other times I would imagine there was more to his reasons for letting me live. Perhaps he enjoyed my piano playing. Perhaps my personality impressed him, my willingness to stare death in the face and proceed anyway. Or perhaps, perhaps, he was attracted to me.

It was easy to find riddle messages to support my wildest speculations. The word "reborn" might seem to be indicated in a riddle, and I would start to find all kinds of messages about reincarnation. I would find "vampire" and "voodoo" and all kinds of things. It was easy enough to find messages indicating that Azazel found me attractive. I was prepared to submit to him, sight unseen, regardless of whether he was a cyclops, a centaur, or a Sasquatch, whether he had horns and a tail or whatever, and whether he would be monogamous or not. If I was willing to risk death to try to save the world, there was no reason to reject an unusual relationship. The monarch should get what he wants, romantically speaking, as the first husband of Camilla Parker Bowles quite sensibly recognized. He readily and gallantly surrendered his wife to Prince Charles as a service to the country.

In a story about lovers warming up to each other in marriage (50) Azazel uses the term "cold zybbuls." In this term I can see this message: "U could bully Oz, boss Oz." This seems to describe the way some men enjoy behavior that is a little bit aggressive or abusive. A tyrant might find it exciting to have his wife lord it over him as a prelude to lovemaking. In "flaxated his modinem" (60), a phrase used by Azazel, several messages are possible, including this: "Dominate me, Leedi, and I shall hold off on an axis shift." There are a couple of other possible messages along this line. Their only real importance to my overall case is that they offer an answer to those who may think that what I am saying must be wrong, because I would have been killed if I had been right.

In "Vainamoinen Glitz, Jr." (51) Asimov tells us: "I am in jail mentalli. Azazel iz never going to let me go." The letters for "jail" occur together in "Elijah Baley" (92). When "Jezebel Navodny," his wife's name, is added to his, the message may be: "I be in jail. I have been jailed by Azazel, hoo iz loony." We have seen a related message in "Hubert Mandel" (98), as follows: "Tu rule the Earth U need tu learn tu bend and tear the human brane." Likewise, in "Bonky Latourette" (107) we can see: "Try not to let your brane be broken by our ruler." This character is under mental stress.

Azazel seems to be linked to messages about reincarnation. In "Eto Demerzel, Cleon I" (66) we can find: "Oz iz cloned, iz determined to clone me." In "Tryma Acarnio" (30) we can find "incarnation," but I don't see a definite message. In "Tomasz Brigon, Gladovia" from "The Nations in Space" (35) the message may be: "I, Asimov, am not glad Oz did damag to mi brain." Brigon has a partner. In "Hamish Mansa, Gladovia" we can find: "I, Asimov, am glad I hav had a high mission." This is similar to the message we found in "Hamish Sankov" (111): "I, Isaak Asimov, hav a hi mission." Mansa and Brigon have a mission high in the sky relating to the world's energy supply. In "the Amazing Larri" from "The Cross of Lorraine" (25) we can find: "Azazel iz harming the Earth." Carl Sagan may have expressed a very personal fear of Azazel in the name "Timofei Gotsridze" (148), where we can easily see "got rid of." The message may be: "I fear it is time for Oz to get rid of me." Sagan lived ten years after the publication of his novel. In "President Lasker, Seymour" (148) Sagan may have been suggesting a message using the words "ask our president," as follows: "Ask U.S. president to kontakt our planet's ruler, a monkey-man." In other messages Sagan seems to caution that any such contact is risky.

Isaac Asimov portrays a plan to decimate the human race in "The Dim Rumble" (14). In the perpetrator's name, "Hannibal West," Asimov may be telling us: "This battle with Asasel is winnable." I pray he is right.

12.
DID EXPLORERS LIE ABOUT SEEING EARTH'S INNER SUN, OR EVEN VISITING THE INTERIOR?

When an explorer makes an extraordinary claim it is natural for us to be a little bit skeptical. Tall tales have been told about sightings of the Loch Ness monster or other fantastic sea creatures. Journalists have lied about coming under fire in a war zone, to make themselves look heroic. Politicians exaggerate their accomplishments. So why should explorers get a free pass to tell us there are holes at the poles?

Danish explorer Fridtjof Nansen may have realized he would have a credibility problem after he and his crew spotted a dull-red sun in the far north at a time of winter darkness in 1894. He may have realized it would be futile to announce to the world that there is a sun at the planetary center. This might have resulted in derision and damage to his career. What he did instead was quite ingenious. He calmly recorded the incident in his diary and dismissed the solar phenomenon as a mirage.

Marshall Gardner gives this account (134):

When the party reached 80 degrees, 1 minute, a remarkable observation was made which may be explained in more than one way:

"…about midday we saw the sun, or, to be more correct, an image of the sun, for it was only a mirage. A peculiar impression was produced by the sight of that glowing fire lit just above the outermost edge of the ice. According to the enthusiastic descriptions given by many Arctic travelers of the first appearance of this god of life after the long winter night, the impression ought to be one of jubilant excitement; but it was not so in my case. We had not expected to see it for some days yet, so that my feeling was rather one of pain—of disappointment, that we must have drifted farther south than we thought. So it was with pleasure I soon discovered that it could not be the sun itself. The mirage was at first like a flattened-out, glowing red streak upon the horizon; later there were two streaks, the one above the other, with a dark space between; and from the main-top I could see four, or even five, such horizontal lines directly over one another, and all of equal length, as if one could only imagine a square dull-red sun, with horizontal streaks across it."

Now it is quite a question whether the mirage that Nansen saw at this time was a mirage of the sun in our sky or whether it might have been some sort of a reflection of the sun in the interior of the earth. Certainly he was not expecting to see the solar light at that time.

Two or three days later this mirage of whichever sun it might have been was seen again.

MODERN EXPLORER REPORTS INNER EARTH TRIP

Another Danish explorer, Edmund Bork, was reported in February 1992 to have entered the Earth's interior, seen the inner sun and visited with the advanced and peaceful people he met on the trip (130). The headlines on the story announced: "There's a hole in the Pole. Scientist finds tropical paradise inside the Earth. UFOs come from inner space— not outer space."

The editors of the *Sun* (tabloid) were a bit cautious with this story. They treated it as a genuine and truthful account, yet they buried it in the middle of the publication. I would never have known about it except that Michael Callinan, my ex-husband, sent it to me.

Some highlights of the article are as follows:

"'We have found a fabulous world,' says Edmund Bork, who led a team of international explorers through the hole in the Pole last summer. 'It has its own sun, a shallow, warm water sea and lush, tropical vegetation.'....

"The Danish explorer refuses specific details on what he and the five other members of his team found....

"Normally, the hole cannot be seen from the air because of the heavy cloud cover over the North Pole, and because the inhabitants of the hollow Earth keep it covered with electronic 'light screens.'... . But the screens fail from time to time....

"Bork and his associates were elated, but not surprised, to find the Earth is hollow—and inhabited. Belief that the world is hollow is thousands of years old, he explains. 'Legends of a hollow Earth and people who live there go back to the ancient Egyptians and Greeks. Even the American Indians believed people lived under our feet.'

"A number of other men claim to have entered the hole in the North Pole. They include William Shavers, a Navy pilot who crashed at the North Pole during World War II, and a tribe of wandering Eskimos who told a Canadian reporter in 1958 that they had found 'a green land at the top of the world.' Leif Erickson and other Norse sea explorers claimed to have found a 'lush green land' near the Arctic circle. 'Later mapmakers thought they were talking about Greenland, but Greenland is hardly green, much less lush,' Bork says.

"Impressive evidence came with a radio transmission from Rear Admiral Richard E. Byrd, the great American polar explorer, as he flew over the North Pole in 1926. He said he saw below him a land of mountains, forests, green vegetation, lakes and rivers."

Byrd also saw a creature resembling a wooly mammoth, according to Raymond Bernard's account in *The Hollow Earth*.

Byrd did not speak of his trip later in life, and many say that neither the trip nor the radio broadcast about it ever took place.

Bork's expedition was inspired, according to the story, by a composite photograph of the north polar view of Earth as it would look if all sides of the planet were illuminated at once. The appearance of what looks like a dark, well defined hole at the center of this November 1968 photograph has inspired a great deal of speculation. Since no central sun's light can be seen within the "hole," I consider this picture to be of limited value as evidence for a hollow Earth. (See Figure 15.)

Both Bernard and Gardner cite discoveries of warm water, pollen and profusions of wildlife in the far north as evidence of the warming influence of a polar hole. These details are of some value, but they obscure the main point. What is really needed is eyewitness testimony that polar holes and a hollow interior really do exist.

Willis George Emerson's novel *The Smoky God* tells of a voyage by a father and son through the north polar opening and all the way down to the south polar opening where they emerged. They reportedly met with the king of the world and his aides, who were extremely tall. The son was able to make it back to his home in northern Europe, but was treated as a madman for sharing his story. The narrator's experience as a mental patient provides what is likely the real reason why this story is presented in fictionalized form. The author knew that if he presented it as truth, the result could be incarceration in an insane asylum.

The accounts of explorers are in agreement as to the inner sun's appearance. It is dull-red or bronze and frequently appears cloudy or smoky. The smoky appearance of Earth's inner sun is confirmed by at least two of Isaac Asimov's riddles.

In "Maskellnik, Eos" (79) we find: "Small son looks like smoke." In "Hiroko, Smool, Monolee" (32) we can find: "Son shines in hollo, looks more like smoke."

The inner sun's light is reflected in the "water sky" near the polar holes, according to Gardner. This description provides an explanation as to what the speaker through Edgar Cayce was probably referring to when he predicted that "His Light will be seen again in the clouds" around the time of an axis shift. New polar holes created by an axis change, leading to an unusual light reflected in the clouds—this must be the real meaning of Cayce's most enigmatic prediction. This prediction serves as powerful confirmation that the explorers' reports described here were honest and accurate.

13.
WHY DOES NASA GET TO BRUSH ASIDE A CONGRESSIONAL QUERY ON MARS POLAR LIGHT?

The NASA response to Representative Wayne Gilchrest over the question of polar light on the night side of Mars was evasive to say the least. It was a little bit like Lily Tomlin's famous line as she acted the part of a telephone operator:

"We don't care. We don't have to. We're the telephone company." NASA wrote to Rep. Gilchrest of Maryland on August 17, 2001 in response to a report of mine entitled: "New Angle Illuminates Mars Night-Side Light Mystery." In spite of my geometric analysis showing that the polar light-cap could not be on the day side, NASA asserted its right to say nothing on this question of night-side light. My report was treated as a "speculative lay concept" because I interpreted the polar light to be emanating from within the planet. To add insult to injury, NASA felt free to list non-photographic evidence like mass and moment of inertia and allege that "all the data" indicated that Mars is solid, at the same time that no comment whatsoever was offered on the photographic evidence.

Since when do we ignore photographs in astronomy and evaluate nearby planets based only on <u>other</u> evidence? I still feel cheated, and Rep. Gilchrest was cheated.

I don't blame NASA. They were kind enough to produce numerous images of Mars that clearly show polar light on the night side. The retouching done to downplay this fact is so obvious and even clumsy that it screams out the fact that the true nature of Mars is being concealed. So NASA has shown us the truth all the while that they have been forced to deny it.

The following hypothetical conversation between NASA officials will illustrate the situation as I see it. The NASA director is shown here interviewing one of the officials—Jeff Bingham—whose office produced the NASA letter.

> NA: This is a routine review of your handling of the Gilchrest letter regarding Ruth Carr's report on Mars night-side light. You aren't in any trouble, but this review has been ordered because of the sensitivity of the issue.
>
> JB: You mean you did not initiate the review yourself?

NA: Nominally I am in charge here, but that's kind of an illusion as you know.

JB: I didn't actually write the letter. You can see it is signed by Kate Rothman, though you can't read her signature, and her name is not printed underneath.

NA: That's a brilliant touch. The letter is essentially anonymous, and that fact should discourage Rep. Gilchrest from questioning its content.

JB: I don't think he was intensely interested in the question of light on the night side of Mars. He presented the material to us in a routine manner with a mild expression of interest. He was basically just doing his duty to his constituent, since she had gone to a lot of trouble to prepare this report.

NA: Ah yes, I have it before me. It looks like her diagram was defective. She calculated the center of the day side as though it were directly north of the bright spot on the photograph, which introduced an error into her figuring.

JB: Yes, we could have pointed that out, but the error was not great enough to alter the result significantly. The entire diagram is very imprecise. That could have been pointed out as well in our letter.

NA: But you decided not to discuss the photographic evidence at all.

JB: We didn't want to concede that there was any merit to Ms. Carr's idea of trying to calculate the location of the center of the day side. Once we started criticizing her methods, if we did so, we could be asked to come up with improvements on her methods in order to answer that question.

NA: Very good. So you threw out the entire question.

JB: In a way we were taking a chance. Members of Congress can be very persistent when they latch on to an issue.

NA: But you didn't think Rep. Gilchrest would press this particular issue.

JB: Not a chance. Even expressing an interest in it nowadays is on a par with questioning the validity of the Moon landing or advocating for the flat Earth theory. I'm surprised Mr. Gilchrest even had the nerve to bring it up.

NA: Still, this is a good time to go over your plan of action in case Congress ever does question you in a more persistent manner.

JB: Our first requirement would be that the discussion be confidential, with absolutely no publicity. Also the discussion should be limited to a small group of people.

NA: And all those people would be encouraged to forget the discussion afterwards.

JB: Well, yes, but the clouding of minds and arranging accidents, cerebral or otherwise—none of that is our responsibility. The higher-ups who have perfected these techniques have not exactly shared them with us.

NA: No, they keep you—they keep everyone—on a short leash. But just for the sake of argument, suppose you were aggressively questioned by a member of Congress about the night-side light on Mars.

JB: I would tell them all what we told Rep. Gilchrest in the letter. We don't have to comment on speculative lay concepts like the theory that Mars is hollow with an internal sun. This kind of a case should be submitted to a scientific journal and let them deal with it.

NA: What if the member of Congress got indignant and said, "I asked you a question, and I don't have to formulate a case and submit it to *Scientific American* in order to get my question answered."

JB: I would say we do not have ready, off the-cuff answers to offer on every question, especially loaded questions related to the hollow planet theory.

NA: What if the member of Congress accused NASA of retouching the Mars photos to downplay the polar illumination?

JB: I would say NASA cannot be blamed for someone else's tinkering with the photographs.

NA: What if the member of Congress said that government has a responsibility to deal with the night-side light question on Mars because it has implications for a future cataclysm?

JB: I would say I was not aware of any such implications.

NA: You would say that under oath?

JB: I would have no choice.

NA: Well, you are very unlikely to be backed into a corner in this way. But if it did happen, you would deal with the situation as well as could be expected. I am satisfied. Thank you very much. (Interview ends.)

Mr. Bingham appears here to be prepared for all challenges on the Mars night-side light issue. Now let us suppose that Rep. Gilchrest stops by his office and actually does question him.

WG: Pardon me, Mr. Bingham. I have a question regarding your office's letter on the matter of light on the night side of Mars.

JB: I did not write that letter.

WG: No, but here it is. Your letter says NASA does not respond to speculative lay concepts and theories proposed by members of the public. My question is, given a brilliant polar "cap" on the night side, how is it speculation to say a sun produced that light? Is that not simply a self-evident fact?

JB: This office does not differentiate between speculation and self-evident fact.

WG: Now, if Ms. Carr had proposed that the polar cap was illuminated by billions of fireflies, I can see where you could accuse her of wild speculation.

JB: We weren't accusing her. We were simply deferring the question.

WG: But you do know the answer.

JB: I cannot say.

WG: You mean to tell me no one at NASA has been the least bit curious about a polar cap's worth of unexplained light?

JB: I haven't asked them.

WG: Would you have provided an answer if Ms. Carr had not offered any opinion as to the source of that light?

JB: It would still have been a loaded question, aimed at supporting a theory that is best dealt with by scientific journals.

WG: So you only answer questions that are asked in the proper spirit, with no evidence of an agenda you disapprove of?

JB: You'll have to excuse me. (Interview ends.)

In real life, of course, Mr. Bingham would not have tolerated so many questions from Rep. Gilchrest, and the congressman would not have been confident enough or bold enough to ask them. He would have left Mr. Bingham's office as puzzled as when he had entered, after being given the runaround.

This is how the question of a secret star in Earth's back yard has been handled. The nearest star to Earth, whose light is plainly seen by us, is the subject of bureaucratic doubletalk and total evasion, as if we had no right to know about this deadly threat to our own planet.

14.
WHY DOES SCIENTIFIC AMERICAN GET TO PRESENT FIELD REVERSAL SCENARIO WITH NO CATACLYSM?

It took tremendous courage for *Scientific American* to publish "The Evolution of Earth's Magnetic Field" in December 1989. The authors, Jeremy Bloxam and David Gubbins, presented extensive evidence that the planet is entering a phase of rapid progress toward a geomagnetic field reversal.

This rare seismic event has been described repeatedly by various authors as some sort of cataclysm, based on evidence of past die-offs. (Rare seismic events can safely be said to be cataclysmic in every case.) The issue is obviously extremely sensitive, since people are easily excited by any news that doomsday is around the corner. An article like this could have set off a firestorm if it had contained a hint that this was a planet-wide cataclysm that was said to be on the horizon. But there was no such hint.

I am sure *Scientific American* was risking not only a public outcry but also the heavy hand of suppression and retaliation by the planet's secret rulers in the interior. Proving that this threat hangs over scientists has been my purpose throughout this book. We recall that the geologist who declared that Edgar Cayce's axis shift prophecy could be fulfilled by an approaching geomagnetic field reversal insisted on remaining anonymous. Most of his credibility was thus surrendered, evidently in an effort to save his own skin. But we don't need to rely on him, or even on the riddles of Isaac Asimov and Carl Sagan. We can see with our own eyes that the solar wind must be defining the axis because its force, as it sets off each polar aurora, is <u>centered</u> at the axis. Each aurora rings the axis. If there were land at the axis it would coincide with the axis. The solar wind's force has to be concentrated at the magnetic poles because it is funnelled there by the doughnut-shaped magnetic field. So the location of the magnetic poles determines how the solar wind will define the axis. When the magnetic poles wander to their new positions in opposite hemispheres, the solar wind will have to flow to those new magnetic poles, thus redefining the axis. This is not theory. This is self-evident fact, which must be well known though not well publicized.

A huge double-page spread provided by Bloxam and Gubbins illustrates how core spots, or hot spots, emerge near Indonesia and travel, one by one, past India and on to the southern Atlantic Ocean, where they "are growing in size and strength and moving southward." They all possess mini magnetic fields that are oriented in the opposite direction from the main field. The growth of these hot spots, while the main field continues to lose strength, looks like a clear sign of a coming reversal. The article states cautiously that Gubbins believes "this may be the process that eventually reverses the earth's field." His reticence, and the fact that Bloxam's opinion is left unstated, gives the impression that maybe this unusual seismic event is far off, or else won't amount to much.

In their 2005 novel *Polar Shift,* Clive Cussler and Paul Kemprecos made use of this phenomenon of hot spots gathering in the southern Atlantic Ocean. They didn't seem to have any doubt that a reversal in Earth's field would result from this process. They portrayed a polar shift as a change in the geomagnetic field as opposed to an actual axis shift, though they did anticipate some sort of earth change that I didn't quite understand. They also portrayed the reversal as being hastened by the villain's action in blasting the hot spots with energy. It was another action-packed Cussler novel with a few elements borrowed from *Scientific American*. The title was misleading, I thought, because people tend to equate a polar shift with an axis shift.

Since I never hear about a coming geomagnetic field reversal from normal news sources, I conclude that scientists as well as the media have utterly failed to inform the public that a rare seismic event is on its way. The full consequences of this coming event have also remained completely shrouded in mystery. Gerritt Verschuur's book *Cosmic Catastrophes,* for example, predicts a field reversal could begin as early as 2030, but nothing is said by him about an Earth axis shift.

Let us suppose that Congress decides to ask some questions about the reversing of Earth's field. Let us suppose David Gubbins, co-author of the *Scientific American* article of December 1989, is available for questioning.

Senator: Thank you, Dr. Gubbins, for appearing before this committee. We feel we have a responsibility to learn more about the reversing of Earth's magnetic field, so we will be able to lead on this issue if the need should arise. Is it fair to say this is a seismic event we are talking about?

> DG: The reversal of the geomagnetic field involves the bubbling up of hot spots near Indonesia and their accumulation in the southern Atlantic Ocean, so yes, clearly this is a process driven by internal heat caused by the decay of radioactive materials. This is true of all seismic activity.
>
> Senator: If this is an extremely rare seismic event, would you charaterize it as a cataclysm?
>
> DG: That is a definite possibility. The precise effects are not known, but some species have gone extinct at these times. It has not been settled as to whether every reversal is a cataclysm.
>
> Senator: You and Jeremy Bloxam, in your article of December 1989 on the reversing of Earth's field, did not breathe a word about the possibility of cataclysmic effects. Why not?
>
> DG: We did not think the event was close enough to warrant setting off a firestorm of public panic.

Senator: Were you sure of that?

DG: Fairly sure, Senator. We did not specialize in the study of cataclysmic effects, nor did we feel we had any kind of a mandate to set off a massive political controversy.

Senator: So there has been a vacuum as far as guidance from experts on this topic. I would guess that trillions of dollars will have to be spent to prepare the country for a rare seismic event. For that we should have decades of advance notice.

DG: Yes, sir.

Senator: But you don't know when the reversal will arrive.

DG: It will probably arrive gradually.

Senator: Will coastlines be affected? Will Earth's axis be affected? If some areas will be destroyed, we should stop building infrastructure there immediately.

DG: Yes, Senator. We don't have a lot of specifics.

Senator: There may be wars going on in areas that will be destroyed, so the countries are fighting over nothing.

DG: Perhaps so.

Senator: I understand the magnetic poles will be relocated in the opposite hemispheres from before.

DG: Yes, that is why it is called a field reversal.

Senator: Would you have us believe that the planetary axis would remain unchanged after magnetic poles were relocated?

DG: There has been speculation as to a change. We don't like to jump to conclusions.

Senator: But you would leave us in the dark regarding such speculation?

DG: Various expert opinions have been published.

Senator: We don't have time to hunt for everyone's opinion and read it. We depend on recognized experts like yourself to keep us informed.

DG: I understand. But I don't have all the answers.

Senator: We feel that scientists have been holding out on us, and we'll be left holding the bag when the world gets a big seismic surprise.

DG: Dr. Bloxam and I have provided a wealth of information in our article. Surely we have been more forthcoming than most experts.

Senator: So we should just be satisfied with what you see fit to tell us.

DG: We can't tell you what we don't know, Senator.

Senator: Thank you, Dr. Gubbins.

The expert has been portrayed here as defending his failure to speak out about a major seismic event on the horizon, based on a lack of knowledge of specific details. But the lack of knowledge of fine details of a cataclysm is not a valid excuse for keeping the public and Congress ignorant of the basic facts of the situation. Especially when scientists receive government support, they should be proactive in sharing what they know about looming catastrophes with political leaders.

15.
CARL SAGAN ON UFO RESCUERS AND REDEEMERS; "ISAAC A. AND I LISTEN AND LOOK IN ON YOU, LEEDY..."

Did Carl Sagan listen to me and watch me in the way Isaac Asimov must have done, based on the evidence in Chapter 8? I have found a riddle that points to that conclusion, but it would be helpful to have more confirmation.

It would have been natural for Asimov to include Sagan in his surveillance program because I was analyzing cataclysmic hints from both of them. I was initially suspicious of Carl Sagan because of his rather scathing criticism of Immanuel Velikovsky, which is not to say I believed Velikovsky's ideas about the "comet" Venus any more than Sagan did. By the time of my 1983 book I had accepted Sagan as a friend of the hollow Earth theory and a coming axis shift. Asimov knew Sagan was his ally in delivering covert cataclysmic messages, and that I recognized this. And so why would Asimov have excluded Sagan from his program of observing me?

In "Nicholas Polydemos," from Sagan's novel *Contact*, we can find this message:

"Isaac A. and I lissen and looc in on yoo, Leedy. I hope yoo don' mind." The spelling here could be better. The strongest point in favor of this message is that the letters for "Leedy," spelled correctly, are grouped together. One might ask whether Sagan provided any riddles based on information he could have gained through surveillance of me. I have noticed one definite possibility, which I wasn't going to mention to spare my ex-husband's feelings. But Michael knows he wasn't doing much to support us during our marriage. He was trying to develop his voice so as to make money from singing, and he also admired his father for many things, including an ability to gain favors from women. He worked more after our marriage was over, and then he moved in with his mother in Florida. In "Michael Kitz" from Sagan's novel, one can find this message:

"I take it that Michael C. iz lazi." I also have wondered whether "Nguyen 'Bobby' Bui" from Sagan's novel referred to a Bobby on whom I had a crush in high school. I thought he was attracted to me, but the signs I was interpreting as attraction were probably in reality signs of embarrassment over a misunderstanding which took place in class. The teacher in the class I attended with Bobby had pressed me (twice) to write a skit for a school production. I wrote a short skit in which she was to wind up being honored with a presentation of flowers. I gave this to Bob, who headed up the Sunday school class for which I played piano. He was a co-leader of this "Mardi Gras" production. I think he interpreted this skit as a sign that I thought there might be sparks flying between him and the teacher. Perhaps she was attracted to him. I didn't know anything about teacher-student relationships in those days, when I was 16. Anyhow, his unusual behavior toward me (blushing, etc.) subsided, but I was becoming quite aroused by this. I was sure we would get together eventually, but it never happened. The skit never happened either. At one point Bob said he didn't have time to deal with it. I suppose he thought the teacher's request for a skit was unfair, which it was. In "Nguyen 'Bobby' Bui" from Sagan's novel, the message may be:

"You nu Bobby be eyeing you." This looks like Sagan might have been saying my crush on Bobby, which everyone in my family knew about, was not based on pure fantasy, but that I actually had a reason for thinking he had been giving me the eye, or flirting. Yes I had a reason, but being young I had misread the situation.

If Sagan found out about this crush I had had, it would not have been through surveillance in the 1980s. It would have been through interviewing people who were familiar with my behavior around 1963, long before I wrote any hollow Earth books.

WAS CARL SAGAN COMMENTING ON UFOS?

A strange collection of letters from Sagan's novel seems to invite analysis as a riddle. One of his characters asks about: "Fobuserdeemus." This is a parody of the sound of "Phobos or Diemos," the names of the moons of Mars. We can find "UFOs" in "Fobuserdeemus." The complete message may be: "Some UFOs R redeemers or frends. Some R bums, murderers." This could be a response to my comments in my books to the effect that people should not believe promises that UFOs are going to rescue the good people from a coming axis shift. My view was that the space people would wind up killing the people who had accepted the rescue offer—killing them and using their blood and organs for biological purposes. For example, the youth hormone interferon can be distilled from large quantities of blood. I like to think Sagan was offering support for what I said about a promised UFO rescue mission, but there is no way to be sure about that. At any rate, I was dismayed at seeing full-page ads in UFO-related literature, promising that the good people would be rescued by space ships at the time of a coming Earth axis shift. I found the same message in a little booklet entitled "We Met the Space People."

MORE ON SAGAN'S "EARTH AXIS SHIFT" RIDDLE

To find that "Earth axis shift" can be spelled from a fictional character's name is astounding enough by itself. But now that you have seen how a reversal of Earth's magnetic field sets off an axis shift, you should be shocked to find that Sagan gave you a character name that can spell "field reversal" <u>in addition</u> to spelling "Earth axis shift." This is a gift of incalculable value, since Sagan's scientific credentials were excellent.

"Stefan Alexeivich Baruda" may contain the following message, as we have seen: "A field reversal shall cause the Earth's axis tu be shifted hi the sun in five tu seven decades." Confirmation can be seen in the fact that Baruda actually does give a doomsday warning in the story. His warning relates to contact from outer space, which has stirred fears of a cataclysm and even an Earth axis shift among members of the public. For Sagan to provide an indirect connection between Baruda and Earth axis shift fears is a truly spectacular warning sign on his part, considering the fact that "Earth axis shift" and "field reversal" can be spelled from the letters in this character's name.

I have mentioned this riddle several times because it is so important and is such an unusual achievement. It is a tour de force which all by itself should cause Sagan to be revered forever.

MORE ON AXIS SHIFTING AND MARS STAR THREAT

Before signing off I would mention a few riddles that somehow were not included earlier, perhaps because they had not been solved. In "Sophronia Latimer" (81) the message may be: "Mars rotation is apt to stall. Then its parts mite hit the Earth." In "Eldridge Latimer" this message can be found: "I am a great riddler. Am I a great leader?" In "Stephen Two, Artemis" (69) Asimov may have been saying: "I hope we steer the Mars star in time so it misses Earth." In "Tapper Savand', Malcomber" (30) some hope for Mars seems to be expressed: "Mars appears to be a doomed planet. Cood Mars be saved?" As we saw earlier, the planet's core would probably have to be re-energized so as to restore the north-south magnetic field, which would allow the solar wind to support the planet's rotation. In "Gottlieb Jones, Marilyn" (6) we see: "Telling yoo Mars rotation is stalling is not going to be an easy job, R.A.L." This realization came to me as a result of the study I did on Earth's rotation in relation to the solar wind. It was a shock to face the fact that Mars had no support for its rotation. I had felt burdened by the axis shift issue, but this new issue was infinitely worse. In "Faraday, Poictesme" (54) we can find: "Mars is poised to come apart forty to fifty years from today." In the same story, "Marcus, Ermentrude" appears to carry this message: "U must understand me. U must steer a star at Mars center." Many solutions are possible for "Genovus Pandaral" (79). Here is one more: "A pole reversal endangers all Europeans." This could be preceded by: "Solve pusles, R.A.L., or else..." In "Lawrence Pentili, Stepan" (28) we see: "Scientists warn new planetari acsis is near. Will people listen?" In "Sheerin, Liliath" (120) we find: "I learn that the Earth shall tear as it tilts." In "Ferdinand Feghoot, Lockmania" (35) the message may be: "I am not a free man. Mi mind locked, I fight to gain control." In "Robert L. Forward, Hughes" (35) we find: "Ruth L., U are free to be a great leader, to tell us the Earth be hollow." In "Manfred Dunkel, Absalom Gelb" from "To Your Health" (49) we find: "U are fearless, R.A.L., and U are a brane. U kould dekode a large number of messages for a book." From this same story, in "Absalom Gelb, Euterpe Weiss" a related message can be found: "People will be able to see mi messages are true."

In "Firebrenner, Atkins, Glennfusion" from "The Instability" (35) the message may be: "If a brain is breaking, konfusion inkreases until seniliti sets in." In "Menander Block" (13) we find: "A monke-man rooler can command me, cood break brane." In "Hilderic, Micrometrica" (56) the message may be: "I came to America a child. I am a riter here at a critical time."

In "Enrico Pavolini" (20) we find: "R.A.L. can pla veri niceli on a piano."

Pavolini's fellow character "Emmanuel Jelinsky" may carry a related message: "Munkey-men like majikal musik." In "Liono Kodell (34) we see: "I like lookin' in on Leedi." In "Berto Noss, Gladia Delmarre" (79) we find: "I am embarrassed to admit I am listening to R.A. Leedi." In "Albert Silverstein" (71) we see: "Brain is in tatters. Still I be alive." In "Benjamin Manfred, Mr. Lineweaver" from "Triple Devil" (67) we can see: "Mi brain be veri badli damaged. I nearli died. Ewe are my lifeline, R.A.L." In "Leonard Koenig" (69) we find: "R.A. Leedi, Karl and I are all in danger. Rooler kan kill anione." In "Civ Novker, Hari Seldon" (30) we find: "I, Isaac A., have never risked rooler's ire as R.A. Leedi has done," In "Merwin Mansky" (121) the message could be: "We risk sew many may win, R.A." In "Hari 'Raven' Seldon" (30) we can see: "Everi riddle I have devised is an SOS sine. I need aid." An alternate solution would be: "I are no savior. I have hardli saved anione." In "Galdo, Marpin, Stinnupad, Shelbik, Numoin, Siferra" (120) the message may be: "Mani millions of people in upper Europe are going to be in danger due to an Earth aksis shift."

The seismic activity associated with a shift in polar openings and in the equatorial bulge may be described in "Chekktor, Stanta" (120), as follows: "Krakatoas are soon to create chaos across the Earth." In "Andrew Harlan, Hobbe Finge" (107) we see: "If brain had not been hobbled I wood warn hollow world of danger." In "Andrew Harlan, Laban Twissell" (107) we see: "I will warn ewe as best I be able. Brain has been twisted bi Asasel." In "Bayta Darell, Toran" (33) we can find: "Eye be not able to tell all. Brane be bent and torn." In "Corporal Quinber" (30) we see: "Ruler ca probe or braque ani brain." In "Sarton Bander, Fallom, Jemby" (32) the message may be: "Moon may be jarred or destroyed by fallen Mars star." In "Hestelonia, Wye, Ziggoreth, North Damiano" (66) we can find: "Azazel has hinted that a star in hollow Mars is going to destroy the Earth." In "Mother Rittah, Smoodgie" (30) we see: "Mars is doomed. Its asteroids or star might hit Earth."

REVISITING ONE OF THE TWIN "PLANET EARTH BE HOLLOW" RIDDLES

An alternate solution to "Hortense Hepzibah Lowot" (45), which is one of Asimov's strongest hollow planet statements, is as follows: "Azazel won't let sientists tell people that planet Earth is hollow. We are onli able to hint at this." An alternate solution to "Merrilee Augina Blankowitz" (63) is as follows: "I know U will be able to learn rulez, win game, R.A.L." In "Arkady Vissarionovich Dezhnev" (29) the message may be: "Is R.A. considered a visionary and a savior? Or is she seen as odd, or sick, or even crazy?" In "another zyltchik" (55) we see: "Yoo look crazy to others, Rooth Anne L., yet the hollo Earth theory iz not crazy." In "Graycloud Five" (66) the message may be: "R.A. Leedy, you could give your life for all." In "Daneel Olivaw" (92) we find: "Do even riddle over and over and ewe will learn." In "Han Fastolfe, Vasilia Aliena" (114) we can find: "The alien sivilisation of hell is not a total fantasi."

Solving a long riddle like this is a challenge, but this is an important riddle because the word "alien" is indicated. In "Theor Realo, Dorlis" from "Death Sentence" (99) we see: "Asasel is lord o'er this hollo Earth." In "Christopher Levan, Emma" (20) we see: "I hope Christ comes in time to save millions ov people phrom a polar shipht." In "Haskell Pritchard" (24) we find: "Ere siks decades pass, the Earth's aksis shall shipht." In "Herbert Bassoon" (116) we find: "The Earth's soon to be reborn." In Bassoon's fellow character "Horatio Trombone" we see: "Inner Earth to be too hot to be a home." This may refer to heat release from seismic activity during an axis shift. In "Hortense, Delancey, Leroy, Rambo" (77) the message appears to be:

"Rooth Anne Leedy's brother Roy Charles Leedy shortens name, becomes Rollean."

In "Arthur Gardner" (25) we can find: "Tu read a great Earth truth, read Gardner." In "Margaret Anne Crawford, Reuters" (122) we may find as follows: "Mi messages are true as to future meteors from Mars and a sun from Mars center. U are warned." In "James Strauss" (103) we can find: "U must trust me. Messajes U see are true." Likewise, in "Elgin Marron" (66) we find: "I am no moron, am a man o' learning." Also along this line, in "Mario Tanto" (66) we see: "Am not a total moron."

In a mood of discouragement, in "Jamin Oser" (79) Asimov appears to have said: "Mi messajes no are seen." In "Biron 'Malaine' Farrill" (115) we find: "I no am a normal, free man. Mi brain be malleable." In "Gillbret oth Hinriad" (115) we see: "Onli bi hinting or in a riddle do I get to tell anione that the Earth be hollo." In "Anselm haut Rodric" (31) we find: "I cannot tell U directli that Mars and Earth are hollo." In "Anselmo Quemot" (113) the issue of racial superiority seems to come up, as follows: "Am not equal to a monqué-man, or so Asasel tells me." In "Henry Jarlow, Blanc and Nkrumah" (63) the message may be: "A hollow world can crack and crumble, can be jarred by a human." An alternate solution would be:

"R.A. Leedy, you know how a larje world can be hollow. You can redeem human race."

In "Vandevanter Robinson" (16) we can find: "I tri to be brave as brain is bent, torn and invaded." A follow-up to this seems to be present in "Abram Ivanov" (39). Through this name Asimov appears to be saying: "I am not vari bray."

That is simply not true. Asimov was a gift that kept on giving decade after decade, giving with courage and determination. Unlike most authors he could reach millions of people, both with stories and with reliable scientific information. I hope to inspire people to love his work and honor him more than ever.

16.

"YOU ARE AS BRAINY AS ANYBODY, R.A." ISAAC ASIMOV EXPLAINS EARTH'S AXIS AND ROTATION.

Proving that Isaac Asimov secretly endorsed my hollow Earth and polar shift literature has not been easy. How could he make a subtle statement that would at the same time be strong and convincing? The Ranay D'Aubisson riddle was probably one of his best efforts.

In "Ranay D'Aubisson" (63) we can find: "You ar as brainy as anybody, R.A."

What makes this a strong statement is the fact that D'Aubisson's main role in the novel *Nemesis* is to provide a brain scan for the heroine Marlene. In other words, the character's profession, which relates to brains, serves to confirm the fact that the word "brain" is an important part of the message in her name. Another important point about this riddle is that it includes all the letters necessary to spell the word "you." It is unusual to find all three of these letters in a riddle, and so it is likely that the word "you" is part of the solution.

Taking this riddle by itself, there is no strong sign as to who the brainy person is who is being praised here-by Asimov. When we find the double "Planet Earth be hollow" riddles as well as other hollow planet messages which Asimov produced, it becomes natural to conclude that the brainy person he was praising was probably a hollow Earth theorist. I had sent him five of my hollow Earth books, three of which had his name in the title. There are so few hollow Earth theorists publishing at any given time that if Asimov was secretly praising a hollow Earth theorist, it would almost have to have been Ruth A. Leedy that he was praising.

We should take a look at several more of Asimov's riddles containing the word "axis," beyond those we have already seen. In "Alexander Hoskins" (36) the message may be: "Dekode and share riddles on noo axis hole." In "Simon Alexander" (27) Asimov may have been saying: "Mani millions need read mi riddles on a noo axis." In "Rashelle of Wye" (66) and her father's name "Mannix IV" the message may be: "I Asimov, say new axis holes will form as axis moves in a reversal."

We have already looked at a message from "Agis XIV" and his friend "Hari Seldon." Looking at "Agis XIV" (30) standing alone, we should be struck by how close this name comes to looking and sounding like the word "axis." Transposing just two letters we can come up with "axis giv." Perhaps the name is supposed to be saying "Axis sags, givs," meaning that the Earth's axis is becoming less well defined in the years leading up to an axis shift.

ASIMOV POINTS TO SUN'S CONTROL OVER THE EARTH

In a name where the letter "x" appears twice, Asimov has a golden opportunity to make a statement about Earth's axis.

"Artaxerxes Schnell" from "The Fights of Spring" (5) needs to have his father's name "Antiochus" added to his own for the names to spell "axis." Then the question becomes, what does one do with the other "x?" This riddle stumped me for a long time, but the meaning is now clear. Let us tackle this important riddle step by step.

1. I note that "Antiochus" begins with five letters that spell "ation," which is the ending to the word "rotation." So I take "rotation" from the two names. That leaves "chusaaxerxes-schnell."
2. To use one "x" I make "axis," reusing the letter "i." This leaves "chuaerxesschnell."
3. The word "Earth" should go with axis and rotation, so I make that word. Actually I make "Earth's," leaving "cuxeschnell."
4. To use the second "x" and two l's, I make "excellent." This leaves "cush."
5. I suspect at this point that Asimov may want to tell us that the sun (or solar wind) controls both the axis and rotation rate of the Earth, as we have discussed. So I take out "sun" and "has control o'er." No letters are left, and the solution is as follows:

"The sun has excellent control o'er the Earth's axis an' rotation rate." There is another riddle in this story that helps to confirm that Asimov is talking about the Earth's axis here. In "Bullwhip Costigan" we find:

"U should all no that the planit's acsis will B shiphting again soon." By itself this is not a strong message because "axis" and "shift" as well as "planet" are all misspelled. But this message and the earlier one about Earth's axis and rotation rate serve to confirm each other and assure us that Asimov has important news for us regarding the Earth's axis. The one remaining character in this story is "Philomel Kribb." She beomes the girlfriend of Artaxerxes, and so we may speculate that "Philomel Kribb, Artaxerxes Schnell" may contain this message:

"I hope my books enable Rooth Anne L. to expose secrets that relate to Earth's hollo shape, its axis an' its rotation rate."

Regarding secrets, here are a couple of my favorite riddles. In "Newton G. Descartes" (57) we find: "Ewe saw dangers were secret and ewe new 'twas wrong, R.A."

Also, in "Edgar Andrev, Ernett Second" (79) the message may be: "R.A. endeavors to end secret dangers." Andrev was in danger as Second secretly prepared to try to assassinate him.

HOW ASIMOV USED A HOLLOW BELL AS AN EARTH SYMBOL

Powerful earthquakes cause the entire planet to "ring" like a bell, as various authors have observed. In other words, the pattern of vibration caused by a powerful earthquake resembles the pattern of vibration caused by the ringing of a bell. Bells, of course, are hollow. A solid Earth might vibrate, but its vibrations would not resemble those caused by the ringing of a hollow bell—probably not.

Isaac Asimov evidently was aware of this phenomenon of the entire planet ringing like a bell. His story "The Singing Bell" (105) portrays Dr. Wendell Urth allowing his own valuable "singing bell," a hollow rocky object that formed naturally on the Moon, to be shattered. Before this happens (as part of an investigation), Dr. Urth strokes the bell, producing music. It seems pretty clear that Dr. Urth's bell is a metaphor for the shape of Earth itself. The breaking of the bell is a metaphor for the damage caused by the migration of polar holes in an axis shift, or even the damage that could be caused by a strike on Earth by pieces of another planet that is getting ready to disintegrate.

The bell of Earth can crack and suffer serious damage. Why else would Dr. Urth's bell be allowed to shatter? As a society we need to wake up and face the music.

The flat Earth theory is accepted by at least ten percent of Americans, even though the theory is vague as to whether Earth is shaped like a pancake or stretches off infinitely in two dimensions. Embracing this theory is just another way of saying that the principles of science and logic don't have to be respected, which of course is true.

The hollow Earth theory has almost no public support, in spite of the quantity of scientific evidence which supports it. We don't want our planet to have the vulnerability that goes with being hollow. We don't want to have to wonder who is living in the interior and how powerful they may be. These issues don't have to be confronted, as long as we exercise our right to believe what makes us feel good. If the truth is to come out, the process will be painful. The end result, however, will be a realization that we need each other to survive, and so an era of increased harmony could be the result.

There are people who have been impressed with the evidence of light on the night side of Mars, but they probably would not have wanted to be quoted on the subject. One expert emphatically did not want to be quoted as having said it was obvious without a diagram that the polar cap on Photo A, or what was said to be a polar cap, was on the night side. A Catholic priest remarked without hesitation that Photo B (inverted) looked like one of those round candle holders with a lighted candle inside (and he had seen plenty of those). A lawyer thought the Mars photographic evidence was worth presenting to our congressman, but he was unable to arrange a meeting between me and Representative Andy Harris or a member of his staff. It is not clear that Rep. Harris would have been able to make any more progress on this issue than his predecessor Waynt Gilchrest had made.

The truth is not hard to see, but it is hard for the truth to prevail against opposition, indifference and fear of the unknown. I am hoping at some point a lot of us will get together and stand up for the truth.

Associated Press

Ultraviolet light photo taken by satellite shows Southern Lights as a ring above the South Pole.

Figure 14. The south polar aurora australis is shown to be a ring of light in this photograph, published February 16, 1985, by *The Morning Call*, Allentown, Pennsylvania.

Scientist finds tropical paradise inside the Earth:

There's a hole in the Pole

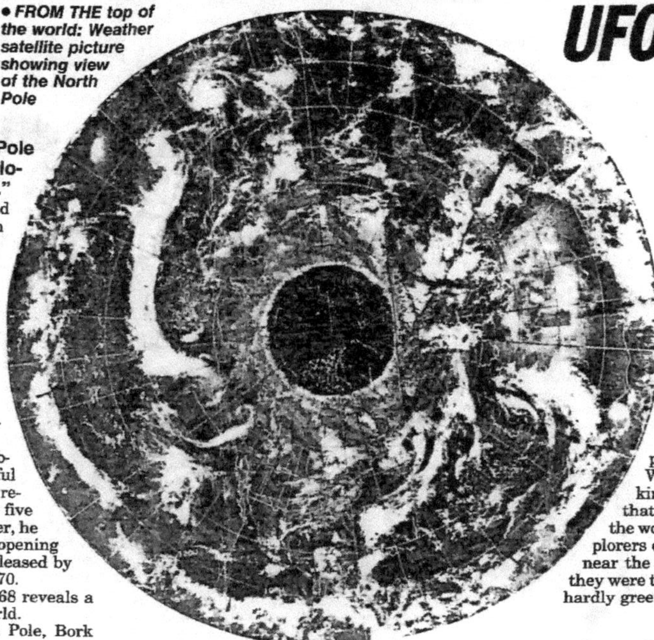

• FROM THE top of the world: Weather satellite picture showing view of the North Pole

UFOs come from inner space – not outer space!

"THERE'S A HOLE in the North Pole and it leads to a tropical paradise located at the center of the Earth."

That's the word from a highly respected scientist and explorer who led a research team to the top of the world last year. The mission was meant to check out astounding satellite photos revealing a large, dark opening at the North Pole.

"We have found a fabulous world," says Edmund Bork, who led a team of international explorers through the hole in the Pole last summer.

Tropical vegetation

"It has its own sun, a shallow, warm water sea and lush, tropical vegetation."

What's more, the land within is inhabited by a highly advanced and very peaceful race of humans. The Danish explorer refuses specific details on what he and the five other members of his team found. However, he says they were led to the 1,400-mile wide opening to the inner earth by a satellite photo released by the U.S. Department of Commerce in 1970.

The photo, taken in late November 1968 reveals a huge, black opening at the top of the world.

There is a similar hole at the South Pole, Bork claims.

"The Earth is shaped like a huge doughnut," he says. "We live on the outer side of the doughnut."

Normally, the hole cannot be seen from the air because of the heavy cloud cover over the North Pole, and because the inhabitants of the hollow Earth keep it covered with electronic "light screens."

The light screens give the illusion of vast fields of ice and snow through holographic manipulation of the real snow and ice that surrounds the opening, says Bork.

But the screens fail from time to time, especially during periods when the sun is in an intensive sunspot phase as it was in 1968.

"The hole is there and all we had to do was walk through the illusion," he says.

Because the hole is so large, the slope down is very gradual and the explorers were hardly aware they were entering another world.

Bork and his associates were elated, but not surprised, to find the Earth is hollow — and inhabited. Belief that the world is hollow is thousands of years old, he explains. "Legends of a hollow Earth and people who live there go back to the ancient Egyptians and Greeks. Even the American Indians believed people lived under our feet."

A number of other men claim to have entered the hole in the North Pole.

They include William Shavers, a Navy pilot who crashed at the North Pole during World War II, and a tribe of wandering Eskimos who told a Canadian reporter in 1958 that they had found "a green land at the top of the world." Leif Erickson and other Norse sea explorers claimed to have found a "lush green land" near the Arctic circle. "Later mapmakers thought they were talking about Greenland, but Greenland is hardly green, much less lush," Bork says.

Impressive evidence

Impressive evidence came with a radio transmission from Rear Admiral Richard E. Byrd, the great American polar explorer, as he flew over the North Pole in 1926.

"He said he saw below him a land of mountains, forests, green vegetation, lakes and rivers," says Bork.

"There was no mention of snow and ice, and the land he described belonged somewhere in the temperate zone, not at the North Pole.

"It sounded a lot like what the Norsemen described and what we discovered."

— JOHN COFFIN

st finds tropical paradise inside the Earth:

ere's a hole in the le

• FROM THE top of the world: Weather satellite picture showing view of the North Pole

UFOs come from inner space – not outer space!

E in the North Pole
›pical paradise lo-
r of the Earth."
a a highly respected
who led a research
orld last year. The
heck out astound-
evealing a large,
orth Pole.
fabulous world,"
no led a team of in-
hrough the hole in

getation

a shallow, warm wa-
pical vegetation."
and within is inhab-
nced and very peaceful
Danish explorer re-
on what he and the five
team found. However, he
he 1,400-mile wide opening
a satellite photo released by
of Commerce in 1970.
late November 1968 reveals a
t the top of the world.
hole at the South Pole, Bork

d like a huge doughnut," he says.
side of the doughnut."
cannot be seen from the air be-
d cover over the North Pole, and
s of the hollow Earth keep it cov-
ight screens."
e the illusion of vast fields of ice
graphic manipulation of the real
ounds the opening, says Bork.
rom time to time, especially dur-

ing periods when the sun is in an intensive sunspot phase as it was in 1968.

"The hole is there and all we had to do was walk through the illusion," he says.

Because the hole is so large, the slope down is very gradual and the explorers were hardly aware they were entering another world.

Bork and his associates were elated, but not surprised, to find the Earth is hollow — and inhabited. Belief that the world is hollow is thousands of years old, he explains. "Legends of a hollow Earth and people who live there go back to the ancient Egyptians and Greeks. Even the American Indians believed people lived under our feet."

A number of other men claim to have entered the hole in the North Pole.

They include William Shavers, a Navy pilot who crashed at the North Pole during World War II, and a tribe of wandering Eskimos who told a Canadian reporter in 1958 that they had found "a green land at the top of the world." Leif Erickson and other Norse sea explorers claimed to have found a "lush green land" near the Arctic circle. "Later mapmakers thought they were talking about Greenland, but Greenland is hardly green, much less lush," Bork says.

Impressive evidence

Impressive evidence came with a radio transmission from Rear Admiral Richard E. Byrd, the great American polar explorer, as he flew over the North Pole in 1926.

"He said he saw below him a land of mountains, forests, green vegetation, lakes and rivers," says Bork.

"There was no mention of snow and ice, and the land he described belonged somewhere in the temperate zone, not at the North Pole.

"It sounded a lot like what the Norsemen described and what we discovered."

— JOHN COFFIN

Figure 15. A report of an inner Earth voyage was carried on February 25, 1992 by *The Sun*, a tabloid news magazine.

Figure 16. A model of a cloudless Venus, as pictured in *Life* magazine, November 1980, is shown on the front cover of my first book, as republished in 1981 by Saucerian Press.

AFTERWORD

When I quoted Edgar Cayce as predicting "there will be trouble" over his cataclysmic revelations, when their meaning is understood, this was done from memory. I recalled Cayce had said there would be trouble, but that those who put their trust in the Lord would not come up missing, but would find much to be thankful for. The idea that there would be trouble was a key point which underscored that important facts had been suppressed. This served to launch the first chapter and really the entire case I was making.

On obtaining a new copy of the Cayce material to replace the copy I no longer possessed (149), I was upset to learn that my quote was slightly inaccurate. The speaker through Cayce had actually made a statement that was somewhat less dramatic and direct. The statement went like this: "Don't think there will not be trouble." This was said in the context of predictions of earth-shaking changes including the shifting of the poles. "These changes on earth will come to pass...." "What is most needed on earth today? That the sons of man be warned that the day of the Lord is near at hand."

When asked when this implied Second Coming would materialize, Cayce said: "When those that are his have made the way clear for Him. Don't think there will not be trouble, but those who put their trust wholly in the Lord will not come up missing, but will find conditions, some way and somehow, much to be thankful for."

I interpret this to mean that those who do the right thing in preparing the public for the cataclysm predicted by Christ will somehow be protected from retribution for exposing secrets that had been carefully guarded. I don't think this refers to putting one's trust in an individual Lord as revealed in the Bible, though such faith and trust may be beneficial. I think this refers to putting one's trust in the importance of standing up for the truth and providing people with needed cataclysmic information.

"To say "There will be trouble" is not much different from saying "Don't think there will not be trouble." I decided not to correct the quote in Chapter 1 because I wanted to avoid the double negative, and did not want to get bogged down at the very beginning with a lengthy explanation of exactly what Cayce was saying. It is, after all, only my own opinion as to what these particular predictions may mean.

For a final look at possible riddles from scientific experts, let us begin with Isaac Asimov's character name "Dr. Jothan Leebig" (113), where the message may be: "The Lord o' the Earth be a big jinn." In "Septimus Johnson, Merceded Gumm" (7) we can find: "Jinns enjoi heering the music U composed, memorised, Ruth." The message in "Juniper Pen, Leander Thomson" (17) may include my name, but not necessarily so. These names may be saying: "Ruth Anne Leedi understands that a major polar hole shipht shall soon destroi northern Europe." An alternate message would be:"U must understand that a major polar hole shipht shall soon destroi northern Europe." In "Melisande Renn, Octavius Ott" from "The Eye of the Beholder" (5), Asimov may have been saying: "I, Isaac Asimov, can listen to U and looc at U, R.A. Leedi." In "Santerestil Mat, Ban" (34) we can find: "Mi brain is in tatters. Asasel is mi master."

Intimidation of scientists by the world's ruler is described dramatically by Carl Sagan in the name "Nadya Rozhdestvenskaya" (148), where the message appears to be: "Oz has everione too skared to try and save the Earth." When the letters for "Earth" are taken from the name, the word "save" is strongly suggested by "sve," and of course the name "Oz" can clearly be seen.

Since "you" can be found in Sagan's character name "Yu Renqiong," I have looked here for a message aimed at me in response to my hollow Earth and polar shift literature. This name from *Contact* may be saying: "You R unique in your inquiring." This may be telling me that I am the only author who has looked at hollow planet disasters both from a scientific perspective and also through the lens of traditional grail symbolism and hints from scientists. At least I hope that is what he may have meant. With the name of Christ showing in his story title "Christmas Without Rodney" (77), Isaac Asimov has created an opportunity to express an opinion about the Second Coming, as follows:

"Ruth Anne understands what Christ meant with his doomsday hints." After casting himself as a solar figure by choosing twelve zodiacal followers (1) Christ was nailed to what has been called a world axis symbol (129), telling us that his promise to return in the clouds refers to a sun at the world axis shining up into the clouds in a new way. He evidently knew the nature of the coming apocalypse could and would be kept secret, and his prophecy could, at the critical time, help us to see the truth and save those most in danger.

NASA has detected a distant planet that is shaped like a collapsed football. This was announced on National Public Radion during the summer of 2024. Why did NASA not explain that this was a stage in the total collapse of a hollow planet that is no longer rotating fast enough to maintain its hollow shape?

Nothing about the hollow planet theory seems to come easily. Since the equatorial region of the planetary shell was thinner than the polar regions (see Figure 13), it makes sense that it would be the first area to go limp. The planet would look like a collapsed football for a period of time before collapsing altogether. Since Mars has lost its north and south magnetic poles. Its rotation longer receives a boost from the solar wind striking those points. We have seen that a message about Mars can be found in "Limmar Ponyets" from Isaac Asimov's novel *Foundation:*

"Mars is not yet (totalli) limp."

BIBLIOGRAPHY

Anderson, Flavia. *The Ancient Secret: In Search of the Holy Grail*; Victor Gollancz, London, 1953.

Angebert, Jean-Michel. *The Occult and the Third Reich*; Macmillan, New York, 1974.

Asimov, Isaac. *A Choice of Catastrophes*; Simon & Schuster, New York, 1979.

——. A Whiff of Death; Doubleday, New York, 1958.

——. *Azazel*; Doubleday, New York, 1988. Stories: 6. "A Matter of Principle;" 7. "Dashing Through the Snow;" 8. "Flight of Fancy;" 9. "Galatea;" 10. "He Travels the Fastest;" 11. "Logic Is Logic;" 12. "More Things in Heaven and Earth;" 13. "Saving Humanity;" 14. "The Dim Rumble;" 15. "The Evil Drink Does;" 16. The Mind's Construction;" 17. "The Two-Centimeter Demon;" 18. "To the Victor;" 19. "Writing Time."

20. ——. *Banquets of the Black Widowers;* Doubleday, New York, 1984. Stories: 21. "A Monday in April;" 22. "Can You Prove It?" 23. "Neither Brute Nor Human;" 24. "The Intruder."

25. ——. *Casebook of the Black Widowers*; Doubleday, New York, 1980. Stories: 26. "Irrelevance!" 27. "The Family Man;" 28. "The Sports Page."

29. ——. *Fantastic Voyage II: Destination Brain*; Doubleday, New York, 1987.

30. ——. *Forward the Foundation;* Doubleday, New York, 1993.

31. ——. *Foundation*; Doubleday, New York, 1951.

32. ——. *Foundation and Earth*; Doubleday, New York, 1986.

33. ——. *Foundation and Empire*; Doubleday, New York, 1952.

34. ——. *Foundation's Edge*; Doubleday, New York, 1982.

35. ——. *Gold: The Final Science Fiction Collection;* HarperCollins, New York, 1995. Stories: 36. "Alexander the God;" 37. "Battle Hymn;" 38. "Cal;" 39. "Fault-Intolerant;" 40. "Frustration;" 41. "Gold;" 42. "Hallucination;" 43. "In the Canyon;" 44. "Kid Brother;" 45. "Perfectly Formal" in "Cal;" 46. "The Introoder" in "Cal;" 47. "The Nations in Space;" 48. "The Shiny Quarter"

in "Cal."

49. ——. *Magic: The Final Fantasy Collection*; HarperCollins, New York, 1996. Stories: 50. "Baby, It's Cold Outside;" 51. "It's a Job;" 52. "March Against the Foe;" 53. "Northwestward;" 54. "Prince Delightful and the Flameless Dragon;" 55. "The Critic on the Hearth;" 56. "The Fable of the Three Princes;" 57. "The Mad Scientist;" 58. "The Time Traveler;" 59. "To Your Health;" 60. "Wine Is a Mocker."

61. ——. *More Tales of the Black Widowers;* Doubleday, New York, 1976. Story: "When No Man Pursueth."

62. ——. *Murder at the ABA*; Doubleday, New York, 1976.

63. ——. *Nemesis*; Doubleday, New York, 1989.

64. ——. *Past, Present and Future;* Prometheus, Buffalo, 1987.

65. ——. *Pebble in the Sky*; Doubleday, New York, 1950.

66. ——. *Prelude to Foundation*; Doubleday, New York, 1988.

67. ——. *Puzzles of the Black Widowers*; Doubleday, New York, 1990. Stories: 68. "Sunset on the Water;" 69. "The Alibi;" 70. "The Fourth Homonym;" 71. "The Lucky Piece;" 72. "The Recipe;" 73. "Unique Is Where You Find It;" 74. "Where Is He?"

75. ——. *Robot Dreams*; Doubleday, New York, 1986. Story: "The Machine that Won the War."

76. ——. *Robot Visions*; Byron Preiss, New York, 1990. Stories: 77. "Christmas Without Rodney;" 78. "Too Bad."

79. ——. *Robots and Empire*; Doubleday, New York, 1985.

80. ——. *Second Foundation*; Doubleday, New York, 1953.

81. ——. *Tales of the Black Widowers*; Doubleday, New York, 1974. Story: "The Obvious Factor."

82. ——. *The Best Science Fiction of Isaac Asimov;* Doubleday, New York, 1986. Stories: 83. "A Loint of Paw;" 84. "Death of a Foy;" 85. "Eyes Do More than See;" 86 "Franchise;" 87. "It's Such a Beautiful Day;" 88. "Spell My Name With an S;" 89. "Strikebreaker;" 90. "The Dead Past;" 91. "Unto the Fourth Generation."

92. ——. *The Caves of Steel;* Doubleday, New York, 1954.

93. ——. *The Complete Stories, Vol. 1;* HarperCollins, New York 1990. Stories: 94. "Gimmicks Three;" 95. "Old-fashioned;" 96. "Profession;" 97. "The Author's Ordeal;" 98. "The Dying Night."

99. ——. *The Complete Stories, Vol. 2;* Doubleday, New York, 1992. Stories: 100. "Lenny;" 101. "Light Verse;" 102. "Mirror Image;" 103. "The Key;" 104. "The Monkey's Finger;" 105. "The Singing Bell;" 106. "The Talking Stone."

107. ——. *The End of Eternity*; Doubleday, New York, 1955.

108. ——. *The Gods Themselves*; Doubleday, New York, 1972.

109. ——. *The Martian Way and Other Stories*; Doubleday, New York, 1955. Stories: 110. "Sucker

Bait;" 111. "The Martian Way;" 112. "Youth."
113. ——. *The Naked Sun;* Doubleday, New York, 1957.
114. ——. *The Robots of Dawn*; Doubleday, New York, 1983.
115. ——. *The Stars Like Dust*; Doubleday, New York, 1951.
116. ——. *The Union Club Mysteries*; Doubleday, New York, 1983. Stories: 117. "Gift;" 118. "Hot or Cold."
119. Asimov, Isaac and Janet Asimov. *The Norby Chronicles*; Walker, New York, 1984.
120. Asimov, Isaac and Robert Silverberg. *Nightfall*; Doubleday, New York, 1990.
121. ——. *The Positronic Man*; Doubleday, New York, 1991.
122. ——. *The Ugly Little Boy*; Doubleday, New York, 1992.
123. Benton, Floria, pseudonym for Ruth Leedy Callinan and Ruth A. Leedy, now Ruth Leedy Carr. *Hollow Earth Apocalypse: Asimov's Warning*; Leedy, Wilmington, Del., 1985.
124. ——. *Hollow Earth Mysteries and the Polar Shift*; Saucerian, Clarksburg, W. Va., 1981.
125. ——. *The Hollow Earth at the End Time*; New Age, Jane Lew, W. Va., 1983. (Also published by Leedy.)
126. Bernard, Raymond. *The Hollow Earth*; Bell, New York, 1969.
127. Blavatsky, Helen P. *The Secret Doctrine;* Theosophical Pub. Co., London, 1888.
128. Bloxam, Jeremy and David Gubbins. "The Evolution of the Earth's Magnetic Field;" *Scientific American*. Vol. 261, No. 6, Dec. 1989.
129. Campbell, Joseph. *The Hero With a Thousand Faces*; Princeton University Press, Princeton, 1949.
130. Coffin, John. "There's a Hole in the Pole;" *The Sun* (tabloid), Feb. 25, 1992.
131. Emerson, Willis George. *The Smoky God: A Voyage to the Inner World*; Palmer, Mundelein, Ill., 1965.
132. *Encyclopedia Britannica*; Benton, Chicago, 1974.
133. Gamow, George. *A Planet Called Earth*; Viking, New York, 1963.
134. Gardner, Marshall B. *A Journey to the Earth's Interior, or Have the Poles Really Been Discovered?*; Aurora, Ill., 1920, reprinted by Health Research, Mokelumne Hill, Cal.
135. Gribbin, John and Stephen Plagemann. *The Jupiter Effect*; Walker, New York, 1974.
136. Heline, Corinne. *Mysteries of the Holy Grail*; New Age, Los Angeles, 1977.
137. Jackson, Joseph. *Pictorial Guide to the Planets*; Crowell, New York, 1973. See title page.
138. Leedy, Ruth A. (See also Benton, Floria.) *Rebirth of a Planet*; Leedy, Wilmington, Del., 1983.
139. *Life*. "Lifting the Veil of Venus;" Vol. 3, No. 11, Nov. 1980.
140. Lowell, John and Kaspar Nureddin. *The Impending Golden Age*; Upland Trails Press, Coolidge, Ariz., 1958.
141. Marvin, Ursula B. *Continental Drift: Evolution of a Concept*; Smithsonian, Washington, 1973.

142. Moffat, Samuel and Elie A. Schneour. *Life Beyond the Earth;* Scholastic Book Services, New York, 1965.

143. Moore, Patrick. *Guide to Mars;* Norton, New York, 1978. (Key photograph omitted in a later edition.)

144. *Mother Earth News*. "Isaac Asimov: Science, Technology and Space;" No. 65, Sept.-Oct. 1980.

145. Page, Thornton and Lou W. Page, Eds. *Space Science and Astronomy*; Macmillan, New York, 1974.

146. Pauwels, Louis and Jacques Bergier. *The Morning of the Magicians*; Stein & Day, New York, 1977.

147. Sagan, Carl. *Broca's Brain;* Random House, New York, 1979.

148. ——. *Contact*; Simon & Schuster, New York, 1985.

149. Stearn, Jess. *Edgar Cayce: The Sleeping Prophet*; Doubleday, New York, 1967.

150. Sullivan, Walter. *Continents in Motion: The New Earth Debate*; McGraw-Hill, New York, 1974.

151. Trench, Brinsley LePoer. *Secret of the Ages: UFOs from Inside the Earth*; Pinnacle, New York, 1974.

152. Verschuur, Gerrit L. *Cosmic Catastrophes*; Addison Wesley, Reading, Mass., 1978.

"I, ASIMOV, am not very good at saving this Earth. Ruth Anne might save it" (p. 17).

"I, SAGAN, have tried to save the Earth, and so has Isaac A., as you write, Ruth Anne" (p. 17).

CLEARCUT CODE FROM ASIMOV PROVES SECRECY MUST EXIST HOW HE MOVED STAR THREAT FROM FICTION TO REALITY

SECRET POLAR LIGHT source revealed by a sketch of Mars. See back, SKETCHES.

An anagram message has deniability. The author can always say no puzzle was intended. But coded messages, once uncovered, are undeniable. Therefore they pose a greater risk for the author, if the secret being revealed is a dangerous secret.

Isaac Asimov was willing to accept this heightened risk for the sake of other people, to save society, to save the Earth. It cuts me to the heart when people take a ho hum attitude toward Asimov's coded messages. Sometimes I think we just don't deserve to be saved. But then, it was ever thus.

"Tar Heel, North Carolina" is discussed in Asimov's story "Irrelevance!" as a possible anagram, and a solution is offered by a character. Did Asimov himself mean this to be an anagram? When I saw that the first five letters could spell "Earth," I knew it must be an anagram from Asimov. But that didn't mean I could solve it. Asimov may have intended this message:

"I cannot teach the hollo Earth theori— the correct theori. I can onli hint at it." One big hint that should not be forgotten is Asimov's use of the term **"hollow world"** in The Robots of Dawn. A thunderstorm's noise was "vibratorily bass in sound when it rattled across a hollow world."

However intriguing a hint or riddle may be, as in these examples, it is still deniable. Even after discovering scores of hints and riddle messages from Asimov and Carl Sagan, I still had it in the back of my mind that I might be imagining the whole thing. And so I was thunderstruck at discovering a real coded message in Asimov's work. The message was definite though the meaning was vague. The message consisted of three five-letter names of female characters from Asimov's cataclysmic

"ICE" ON MARS NIGHT SIDE? ASIMOV'S SCORN MARS A "LOOSE CANNON," STELLAR "WARHEAD"

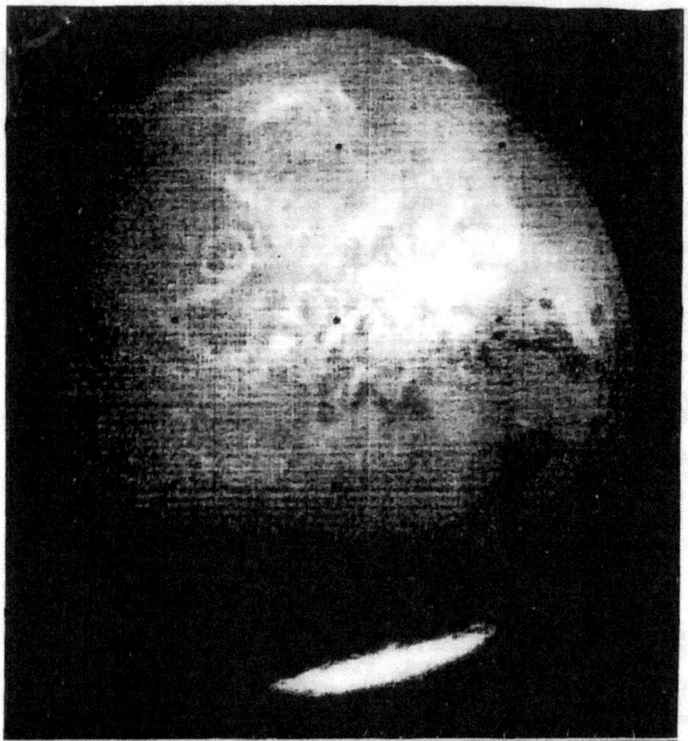

Figure 1. This Mariner 7 photograph of Mars, referred to here as Photo A, is the first in a series of four pictures taken in sequence, all showing an extraordinarily bright south polar "cap" on the night side. This picture is from Space Science and Astronomy (Page) and carries this caption: "Mars from Mariner 7 at 293,200 mi. distance, on Aug. 4, 1969 at 10:38 UT with longitude 115 degrees centered. Nix Olympica (later renamed 'Olympus Mons') is the white circle 3/4 in. toward upper left from center. Bright south-polar cap at bottom. (JPL-NASA photo.)" The planet is in about 3/4 phase here, which would be more obvious if the southern hemisphere were not so dark on the lower right, perhaps artificially darkened.

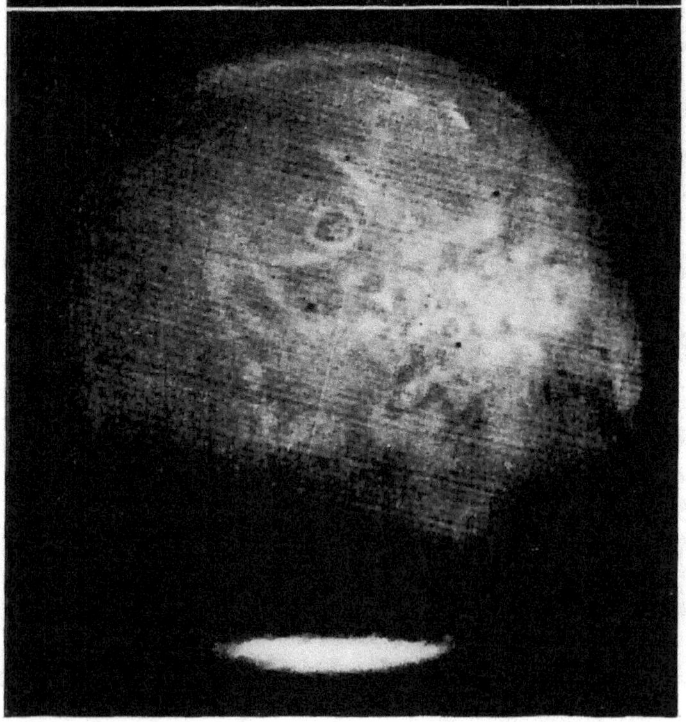

Figure 2. Referred to here as Photo B, this Mariner 7 photograph of Mars, taken shortly after Photo A, is from Patrick Moore's Guide to Mars (1978 edition). Both Olympus Mons and the sun's bright reflection are further east in this picture, which is in about 2/3 phase. A small dot from the dot matrix should be showing near the upper right edge of the polar "cap," but I foolishly whited it out some years ago. The edges of two such dots can be seen on the polar "cap" in Photo D where the lower part was cropped off. It is these dots, superimposed on the polar cap, which prove the polar cap is not some sort of computer error but a genuine part of the image as seen by Mariner 7. This "cap" can only be the light of an inner sun.

Carr- Decret- p. 6.

Sophie" we find:

"Mars shall kollapse, break apart. Its star shall be seen in the sky." In "Mother Rittah" from Forward the Foundation we see:

"A mammoth meteor mite hit the Earth." From the same novel, in "Laskin Joranum" we find:

"A sun in Mars is a jinks on us all in a loos kannon." Other examples along this line will be seen.

Asimov and Sagan Make SOS With Dots.

SOS signs are part of the plot of Isaac Asimov's story "Old-fashioned," where two spacemen are marooned in the asteroid belt. Therefore, when we see numerous sets of three dots in the story we may wonder whether Morse code "S" is being created here. These potential "S" signs (with punctuation added) are grouped in three closely spaced pairs, with only one set of three dots standing alone. When we look in between each pair of three-dot sets we do not find three hyphens or dashes representing the letter "O." Instead we find a word suggesting a circle or letter "O." In one interval the word "orbit" is found. In another the word "hole" is present. In the third interval the word "oh" is used, along with the word "pill."

The purpose of the SOS signs in the story plot is not only to achieve

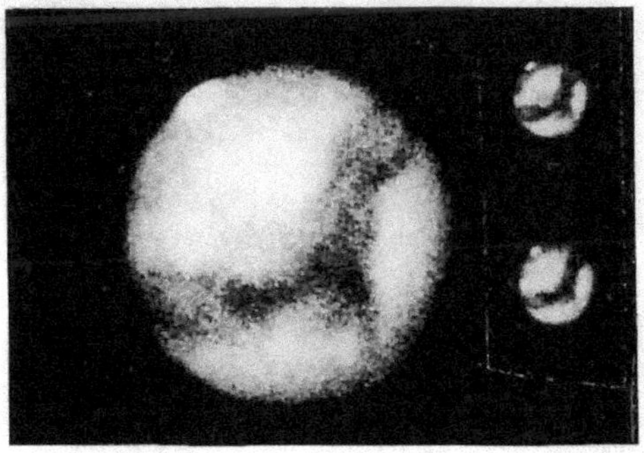

PROTRUDING BLOBS of polar light on these photos of Mars from 1902 and 1909 were for years a challenge for scientists to explain, since there was no definite evidence. George Gamow in his book A Planet Called Earth called the bright spot a polar ice cap. Marshall B. Gardner in his 1920 book identified these spots as the light of an interior sun shining through polar openings. Though Gamow's scientific credentials were better, he was wrong to make a definite statement in the absence of evidence. While he and other scientists were referring to these so-called ice caps, other authorities were frequently wondering whether we would ever discover water on Mars. Wet rocks have now been detected near the poles, not a bright cap of ice.

Figure 3. This Mariner 7 photograph of Mars, taken Aug. 4, 1969, is from the Encyclopedia Britannica, 1974. This Photo C, as we shall call it, shows a slight progression to the right of Olympus Mons and the sun's bright reflection, as compared to their positions in Photo B. Because it was artificially lightened, this photo does not appear nearly as shaded as it should with the sun's illumination decreasing. The darkness of the background sky has been extended to about 3/4 of the circumference of the planet's disc, in spite of the fact that exactly 180 degrees of that outline is illuminated by the sun as a heavenly body goes through phases. Note the clumsy glitch in the outline on the right side. The polar "cap," with its unexplained brightness, has thus been artificially placed on the day side.

Figure 4. This Mariner 7 photograph of Mars, which we will call Photo D, appears on the title page of Pictorial Guide to the Planets (Jackson). As in Photo C, there is artificial definition of the planetary outline extending under the polar "cap," that is, onto the night side. The impropriety of this may explain why most of the "cap" has been cropped off. Though the brightest solar reflection point has disappeared from the right side, this photo has been made to appear as though brightness extends as far to the left as on Photo A. As Mariner 7 has moved toward the dark side of Mars, the polar "cap" is now a somewhat fatter oval of light.

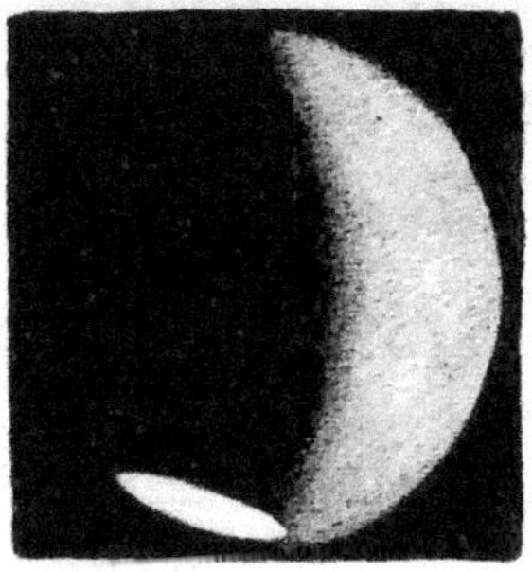

SKETCHES of Photo C (on front) and Photo D (at left) were done to show how those Mariner 7 photos of Mars must have looked prior to extensive retouching that was evidently done. The day-night border at the right edge of the polar "cap," if not concealed, would have revealed that the "cap" was on the night side, and therefore illuminated from within the planet. Measuring 90 degrees from the sun's bright reflection on Photo C, the day-night border on the lower perimeter must be as shown on the sketch.

"OUR WORLD IS HOLLOW; UPPER EUROPE SHALL BE WIPED AWAY BY A POLAR HOLE." — ASIMOV

The above message was issued by Isaac Asimov in two names from a story about a man struggling and winning against a demon. The names "Isidore Wellby, Shapur" from "Gimmicks Three" could generate other messages of course... But how about another demonic story where two mated character names each contain:

"PLANET EARTH BE HOLLOW." The names "Winthrop Carver Cabwell" and "Hortense Hepzibah Lowot" from "Perfectly Formal" in "Cal" are the only two character names in all of Asimov's work that contain all those words. (The latter name can also spell "is.") Do you really think such powerful lightning could accidentally strike twice in the same story, with the names of an engaged couple, and not strike anywhere else? When Winthrop drops Hortense for a more voluptuous woman and dismisses Hortense as "frontally concave," can you not see that Asimov is hinting about the concave inner surface of hollow planet Earth, which is strong confirmation of the message in Hortense's name?

"I STRIVE T' SAVE PLANET." This message can be found in "Stettin Palver," who in Asimov's final novel Forward the Foundation joins a team establishing a new outpost for civilization to be restarted and preserved on a new planet. We will be looking at names like Planchet and Platen from Asimov, which he used to make important statements containing the word "planet."

"Eye R not free." Carl Sagan provided this obvious message in the name of "Tyrone Free" in his novel Contact— a character who is not free because he is a prisoner. By including "free" in the (Cont. on p. 12.)

"AFTER I DO A RIDDLE I FEAR RULER...."

"YOU HOLD A HOLLOW EARTH IN YOUR HAND"

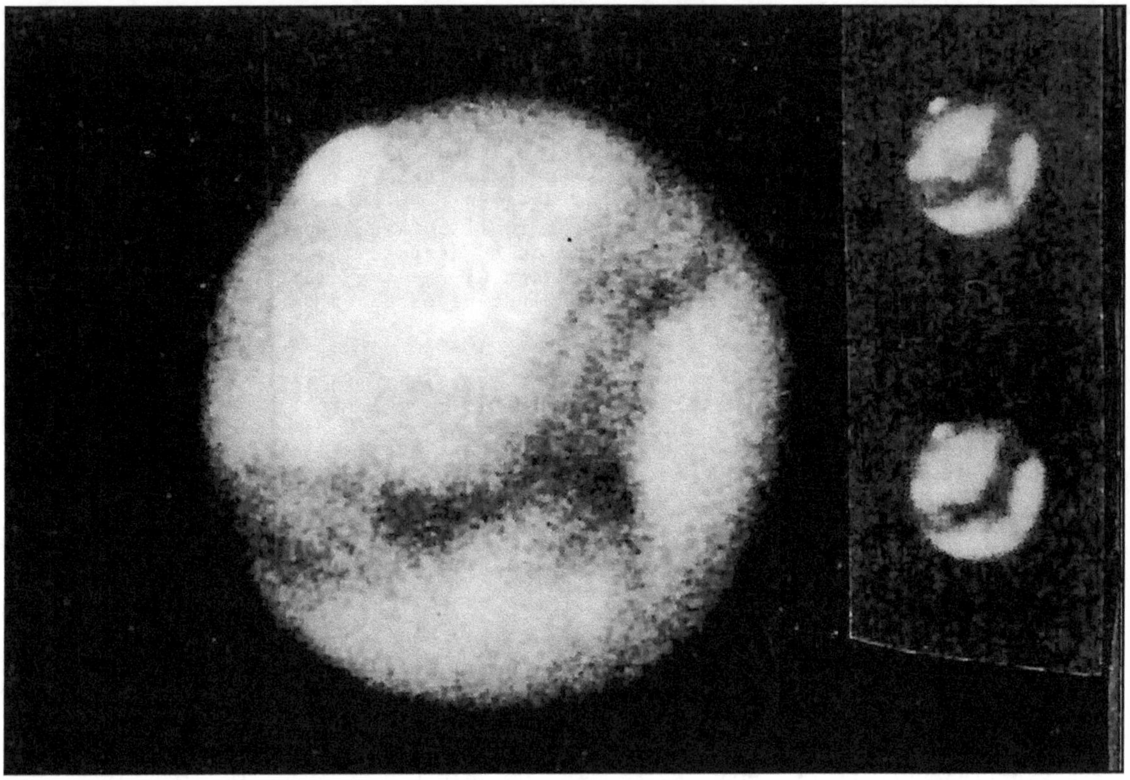

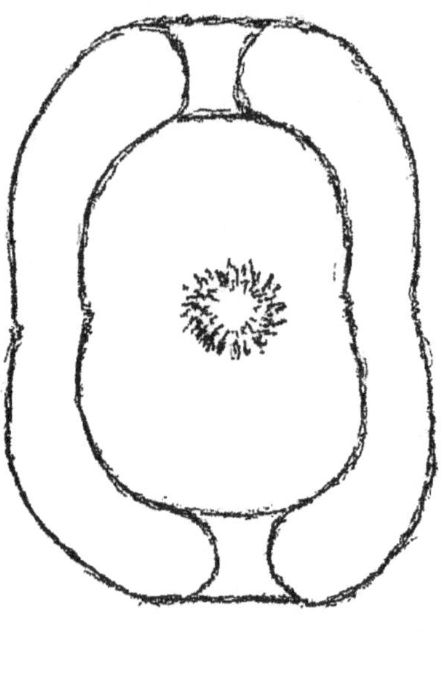

WHY MARS
could be the
next slowly
rotating
planet to
collapse.

See p. 104.

www.ingramcontent.com/pod-product-compliance
Lightning Source LLC
LaVergne TN
LVHW082351200926
848481LV00012B/3533